SOCIAL CONSTRAINTS
ON
TECHNOLOGICAL PROGRESS

Social Constraints on Technological Progress

ROB COPPOCK
International Institute for Environment and Society of the Science Center, Berlin

WZB-Publications
WISSENSCHAFTSZENTRUM BERLIN

Gower

Published by
Gower Publishing Company Limited,
Gower House, Croft Road, Aldershot, Hampshire GU11 3HR, England

and

Gower Publishing Company,
Old Post Road, Brookfield, Vermont 05036, U.S.A.

British Library Cataloguing in Publication Data

Coppock, Rob
Social constraints on technological progress.
1. Technology assessment 2. Risk management
I. Title
363.1 T174.5

ISBN 0-566-00754-1

CONTENTS

ACKNOWLEDGEMENTS

This study originally fulfilled the requirements of a doctoral dissertation in the Economics Department (Fachbereich Wirtschaftswissenschaft) at the University of Wuppertal in West Germany. I am thus deeply indebted to the advice and guidance provided by my advisor, Professor Biernd Biervert, and the second principal reviewer, Professor Wolfgang Sodeur.

I would also like to express my thanks to Professor Timothy O'Riordan, of the University of East Anglia, and Professor Helmut Jungermann, of the Technical University in Berlin, for their helpful comments on various drafts. My gratitude, not only for invaluable comments and suggestions but also for her encouragement and support at crucial points in the writing process, goes to Camilla Krebsbach-Gnath, of Battelle-Institut in Frankfurt.

Thanks also to my colleagues at the International Institute for Environment and Society and the other parts of the Science Center Berlin for discussions too numerous to detail here.

Finally, my appreciation to my wife, Arlette, for her patient understanding of the frustrations accompanying many early morning writing sessions over rather more months than we had anticipated.

Rob Coppock

Berlin, January 1984

1 INTRODUCTION

> The Church welcomes technological progress and receives it with love, for it is an indubitable fact that technological progress comes from God and must lead to him.
>
> Pope Pius XII
> (1876-1958)
> Christmas Message
> 1953

> We are well aware ... that the future of man and mankind is threatened, radically threatened, despite very noble intentions, by men of science. And it is menaced because the tremendous results of their discoveries, especially regarding the natural sciences, have been and continue to be exploited--for ends which have nothing to do with the prerequisites of science, but with the ends of destruction and death ....
>
> Pope John Paul II
> Speech before
> UNESCO
> 2 June 1980

Recent years have revealed significant changes in attitudes about the benefits and risks of scientific and technological progress compared with just a few decades ago. Technologies have been developed and applied to virtually every aspect of human activity: to strengthen political and military power, to raise productivity in the workplace, to lighten the workload in the home, to control the impact of environmental hazards--the list may be extended almost without limit. While the benefits produced by these developments are obviously significant, a large proportion also have been accompanied by unforseen negative side-effects.

Technological hazards present complex problems which often involve diverse groups of individuals and organizations. How to respond to such hazards increasingly seems to be a subject of political as well as technical debate. Growing understanding and public recognition of these effects has shifted the very nature of the discussion about technological progress from exclusive emphasis on the advantages to heightened awareness of the risks that technology entails. The focus of attention of many technological issues is no longer on primarily technical considerations but rather has shifted to an uneasy integration of technical and political considerations.

This is the context within which this study should be seen. The overall aim is to develop a conceptual framework within which the scientific and technical judgments can be placed in proper relation to the political judgments which are required to deal with such issues in a rational and intentional manner (i.e., acting on the basis of deliberate examination of relevant considerations).

The title of the study is "Managing Technological Hazards". In a sense it must define what the act of managing technological hazards entails. Thus, central questions are: Who should act? What should be done? In whose interest? With which legimation? Consideration is given to many more people than merely those who operate the technologies producing the hazard in question.

Possible responses include indirect action relying on various kinds of interdependencies among relevant groups and organizations as well as direct intervention in the chain of events from source to consequence. Reference is made to a variety of groups and sets of individuals whose respective interests are relevant for the operation of a particular technololgy or the control of its hazards. Typical patterns of interaction among such groups and organizations are described, and a set of guidelines for making decisions in such situations developed. The importance of certain prevalent patterns in modern industrial society is also examined. Thus, the underlying problem is how to integrate all these different concerns in a single decision or program of action.

An analysis of articles in the New York Times between 1950 and 1978 illustrates the change over time in the focus of discussion about technological hazards. (Dierkes, 1980, pp. 22-23)

> Briefly, the concern with technological risk in the mid-1950s was limited to nuclear fallout from military testing and to the construction of superhighways. A significant change is noted in the mid-1960s. The majority of the articles remain positive, but the number of articles sceptical of technological developments increases. Articles referring to ecological damages begin to appear, and whereas the sources of information for articles during the previous period were largely confined to the defense industry and government, citizen action groups were now mentioned for the first time as sources of information. Military and economic arguments supporting the introduction of new technologies practically disappear. Although a concern for health and safety is evidenced throughout the whole period of analysis, it is given more prominence only since the mid-1960s when people began to notice the impact of technology on the quality of life as a whole.

Other analyses suggest that this recognition of the negative aspects of technological awareness in the popular press was preceded somewhat by reports in the professional scientific literature. Figure 1.1 illustrates that relevant reports in the journal <u>Science</u> increased substantially a few years before such issues began to appear in major American newspapers. Changes in public opinion, as measured with survey methods, lagged somewhat behind reports in the popular media, but shows a similar pattern of increase. Although the data presented in Figure 1.1 refer to material from the United States, it is generally felt that similar concusions may be drawn for the other parts of the world which have industrialized market economies. In particular, Inglehart (1977) has argued that similar political and ideological developments are to be found in these countries.

In recent years, considerable analysis has been devoted to the examination of factors associated with this increase in professional and public concern about technological hazards. A body of literature has emerged on what is called "technology assessment" (see: Porter et.al. 1980) and "risk analysis" (see: Coppock, 1981). Detailed analyses of specific technologies have

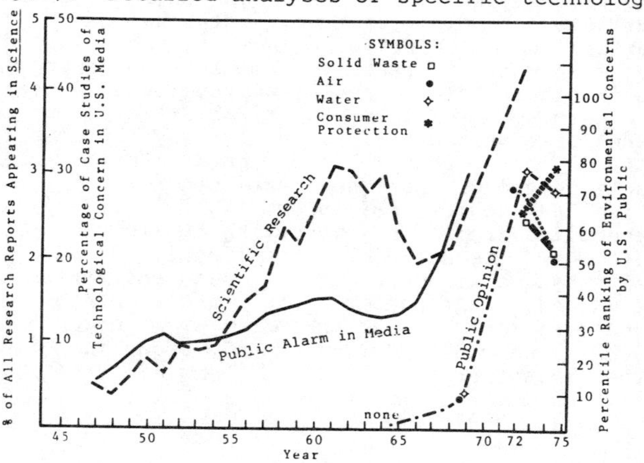

Scientific Research is based on analysis of research reports appearing in the interdisciplinary journal Science (Halverson and Pijawka, 1974). Public Alarm over Technology is based on a 5-year running average of 45 case studies of alarm as evidenced by major newspapers and periodicals (Lawless, 1974). Public Opinion is based on rankings of between 20-30 worries and concerns from repeated nation-wide U.S. opinion polls. In 1964 no environmental concerns were deemed important enough to list.

Figure 1.1 Trends in Scientific Research, Media and Public Opinion

Source: Kates, 1980, p. 80

been performed by various organizations in different countries. Perhaps the largest and best-known example is the Reactor Safety Study of the United States Nuclear Regulatory Commission. This single study involved hundreds of man-years and required nearly five years to complete. But one should also refer to the many studies that have been performed by the U.S. Office of Technology Assessment and by other research organizations. This pattern is duplicated in other industrialized countries, so that the overall effort to improve the understanding of technological hazards and their consequences has been immense.

Yet despite this effort and the resulting improvement in both the analytical capability and in the knowledge about specific hazards and how to respond to them, public alarm seems to have grown even greater. Why is this? It is not merely another manifestation of a public seeking sensational stimulation; it is more than the technological equivalent of tabloid headlines and shocking photos. But this requires careful analysis. Careful analysis because it is not easy to ferret out the complex influences over public opinion on the one hand, and on the other, because if this increased alarm is associated with sober and firmly held opinions rather than unthinking whims, then the consequences for technology planning and decision making can be serious indeed. (Otway & von Winterfeldt, 1982)

An important issue for the management of technology is thus whether technological hazards justify such concern or if such alarm is merely an aberration, somewhat like "noise" in an electric circuit. But justified or not, the mere existence of such fear necessitates dealing with it. A second issue is therefore how to identify the proper response to such fears.

These questions--how to integrate diverse concerns in a

single program, and how to determine proper response to public fears--form the point of departure for this study, the overall aim of which is to contribute towards rational and intentional response to technological hazards.

The Backdrop

It is important at the outset to set the context within which technological hazards should be seen. No attempt will be made to give a precise definition of the term "technology". In general, technology refers the sum of the means employed to provide objects necessary for human sustenance and comfort. Thus, the concept of technology incorporates the existence as well as the use of physical artifacts of some sort. The exact nature of the technology in question, of both the physical objects and the way they are used, is obviously important in determining the nature of the hazards associated with that technology. Defined in this way, it is obvious that the number of possible technologies is virtually unlimited. Indeed, the number of technologies actually employed by mankind is probably innumerable.

Furthermore, while both the nature of the technology and the manner in which it was introduced and is employed can influence the response to the hazards associated with a technology, it is possible to focus on hazards rather than technologies. In fact, laws and regulations often address the negative concequences of technologies and their use rather than the technologies producing them. This may be because the ultimate interest is in avoiding the undesirable consequence, and the existence of multiple sources would require considerable repetition if the focus were on technologies as such.

Considerable work has been done over the last few decades which ostensibly analyses technological hazards and proposes appropriate responses. This has usually led to generalizations about kinds of hazards and responses to them rather than about technologies. It is in this spirit that this study is conceived. The study presents a conceptual framework for the management of technological hazards. The focus is on hazards, and the concepts and ideas are intended to be applicable to a variety of technologies with different attributes.

The development of relevant analytical approaches, and especially the concerns which dominate them, is an important influence on the conceptualization of this study. At the beginning of the period considered in this study (roughly from the 1950s onward), the majority of attention was devoted to the technical aspects of technological hazards. Social progress was widely viewed to be synonomous with technological progress. Somewhat later, as public awareness of the negative side-effects accompanying many technologies increased, attention shifted to personal, psychological reactions to the hazards and their consequences. More recently, attention has turned to the conflicts surrounding the use of hazardous technologies and the attempts to analyze and respond to them.

The unregulated use of technology has proven unacceptable. Asbestos, for example, was produced with relatively few precautions for years after its medical dangers were widely acknowledged. Only when workers began to exhibit clear symptoms in disproportionate numbers, and some to die, were appropriate measures taken. As a larger proportion of the population has become well educated and capable of understanding such matters, the recognition of the extent of technological hazards and of the general inattention to them has grown more widespread. This growing awareness of technological hazards includes two important elements: not only is it generally difficult to recognize and understand technological hazards and their consequences, but there is usually little motivation for those who produce such hazards (often business enterprises) to undertake the effort unless forced to do so. Furthermore, the corporations that usually prove to be the "producers" of technological hazards are most often quite resistant to such pressures. There appeared to be little that the individual, or even groups of citizens banded together, could do. Thus, the problem has been not only one of the existence of technological hazards, but also of the lack of response in the overall industrial system which produces them.

The existence of technological hazards and the difficulty of dealing with them contributes to two quite different ideological responses. (O'Riordan, 1981) On the one hand, some individuals developed what has been called the "ecocentric" point of view, holding that human activity should be organized so as to minimize its negative consequences for both people and the environment in general. On the other hand, those holding the "technocentric" position believe that human activity should be organized so as to make the most efficient use of the resources to be found in the environment. Although these two points of view are otherwise in general disagreement, they both concur on one point: human activity, and especially that involving the application of technology, should be carefully examined. The reasons for performing such analysis are of course different from the two points of view. The ecocentrist wishes to minimize undesirable consequences, the technocentrist to optimize the application of technology. But both require information and analysis. The remarkable progress which has been made since the second world war in technologies for measurement and analysis have therefore been of major importance in the emergence of alarm about technological hazards. Both those wishing to further the use of technology and those attempting to deal with its negative consequences have capitalized upon, for example, improved precision in measurements of various sorts of conditions in the environment and consequences for humans.

One consequence is that we simply have a much better understanding of the functioning of the technologies as well as their human and environmental consequences than used to be the case. (Lowrance, 1976) This increased knowledge contributes to a hightened sense of alarm because it reveals that the extent of technological hazards is much greater than had been recognized previously.

There has been, of course, a tremendous increase in the extent of governmental regulation of techological hazards since the 1950s. Governments typically attempt both to stimulate the development of new technologies and to control the negative consequences of existing technologies to a much greater degree than was previously the case. This has been a significant contribution to the growth of the public sector.

But private firms have also responded to the situation presented by widespread technological hazards. Some firms have recognized the advantages of taking action on their own initiative before government intervention is imposed. This is perhaps clearest in the chemical industry, where some firms strive to identify problems and avoid them before they become public issues. (Hanley, 1981) Many business firms contribute to the activities of industry associations or independent research facilities as an attempt to deal with problems in their branch. (OECD, 1981)

But it would be incorrect to characterize the response to technological hazards as emanating primarily from government and secondarily from business. Private groups and individuals have become quite active with respect to such hazards. In some respects, it is the activities of environmental and other public interest groups that "force" many of the regulatory and industrial responses we can observe. For example, at a meeting in 1979 discussing the perception and handling of technological risk in the European Community, one of the clearest conclusions was the appeal for a better understanding of why some technological issues become political issues while others do not. (Coppock, 1980)

Thus, we can say that the effects of various technologies are felt by many different groups in society. At the same time, it must be recognized that the use of a technology, as well as the resulting products and other consequences, is subject to a variety of influences from many differnt persons, groups, and organizations. The complexity of the relevant patterns of influence is one of the central attributes of the overall problem of managing technological hazards.

A second major attribute is that technological conflicts often appear to hinge not so much on the specific hazards of the particular technology involved as on some broader, usually ill-defined, issue. Some technological conflicts have clearly become political issues. Several explanations for this have been proposed. For example, conflicts about energy production technologies have been attributed by some observers to fundamental differences concerning the aims and purposes of public measures affecting the availability of primary energy. (Wildavsky & Tenenbaum, 1981) Others see the nuclear power debate as reflecting the emergence of a large group of people possessing "new" values which conflict with the "traditional" values associated with technological progress. (Bechmann, Frederichs & Gloede, 1982) But it is also argued that the unavoidable uncertainties involved make technical and scientific conclusions about the risks involved no more than the collective best judgment of a group of experts, and therefore hardly more substantive than are the reactions of the general public, whatever the basis of public attitudes. (Otway & von Winterfeldt, 1982)

Such conflicts can also be seen as part of a much larger and more fundamental development in modern industrial societies. It has been claimed that the basic social conflict in industrial societies is no longer the exploitation of certain social or economic classes but rather the alienation of large parts of the population. (Touraine, 1971) According to this view, the processes of social domination can no longer be viewed in strictly economic terms. Rather, the individual is subjected to forces of all kinds, in every aspect of his life--in education (which has been given over to the state), in the workplace (which is in the hands of his employers), and at home (where activities are largely determined by the specturm of consumer goods and services). The individual is forced to participate, to accept a role in the system which has been defined for him by others. His only alternative is to remove himself from the major interactions of the society, to "drop out" of the typical patterns of social interaction.

It is quite easy to see that the individual might view the imposition of exposure to technological hazards as just another example of this pattern of "dependent participation". He is not consulted about the matter. The decision is made for him by someone in some unknown part of an ill-defined decision structure. Nor is he even informed about what is to happen to him after the decision has been made. If he ever discovers what was decided, much less the factors considered in the deliberation, it is probably because he is able to observe some change in his life or in the environment in which he lives. Thus, it is quite plausible that individual and group response to specific technological hazards might reflect more an attempt to combat the overall decision structure in society than a direct response to the hazards involved in the specific case.

The resulting confusion about exactly what is at issue is exacerbated by a lack of agreement as to terminology. It is fairly common to encounter questions about the acceptability of a particular risk. But the term "risk" is used in many different ways. In economics, risk usually refers to the possibility of incurring profit or loss. The risk involved is the uncertainty of the outcome. But the development of an analytical approach called "benefit-risk analysis" implies something quite different. In this usage, risk refers to the "down side of the gamble", to the potential damage or loss and excludes the potential gain (which by implication is in the benefit side of the comparison). This usage corresponds to the practice of many business firms, where "risk management" is considered to be a part of financial management. Its aim is to optimize the firm's behavior with respect to potential loss-producing events. These examples illustrate that there is not only the confusion about the scope of the relevant preferences (referred to above), but also about exactly what should be examined.

The lack of widely accepted definitions for the various terms not only contributes to confusion, but sometimes also leads to imprecision. Therefore, a single set of definitions will be used throughout this study. Although they are not necessarily in wide use, it is felt worthwile to insure a consistent use of terminology, and to present these terms at the outset.

A "hazard" is a potential event with consequeces which cause damage to some person(s) or thing(s). A hazard carries the connotation of danger, and in some cases fear on the part of the persons exposed to it.

"Exposure" refers to bearing one or more consequences of the potential hazard. This implies the existence of one or more causal pathways, or chains of events, leading from the initiating event (which may be said to present the hazard) to the undesirable effects for the exposed individual or object.

The direct consequences of a hazard are called "effects". Although these are often beneficial as well as damaging or undesirable, there are certain circumstances for which it seems appropriate to give the preponderance of attention to the negative effects. These latter circumstances typically dominate the analyses found in what has been called risk assessment, risk analysis, or risk management.

The likelihood that a hazard will occur combined with the likelihood that it will lead to the negative effects in those exposed to it is the "risk" associated with exposure to that hazard. Although the concept of probability can be formally applied only to specific events in experimental conditions including a fixed time reference, the degree of risk is often expressed in probabilistic terms even though such information is incomplete. Probabilistic expressions of risk in such circumstances are not meaningless, but care must be exercised to interpret them correctly. Risk is also often aggregated across groups, as in references to individual, group and societal risk. Similar care must be exercised in such uses.

The evaluation of the exposure to a hazard, or the assessment of the acceptability of the risk, often involves reference to the "reaction" of exposed persons or of the public in general. This incorporates the way they understand the hazard, its likelihood, and the consequences as well as the way they "feel" about the hazard and its consequences. Thus, reaction combines both rational and emotional aspects, and characterizes the person's state of mind vis a vis the hazard in question. The action undertaken upon learning of exposure to a hazard, or alteration in one's pattern of behavior, is called the "response".

It is difficult to go much beyond these general terms without referring either to specific hazards and consequences or to particular procedures for analyzing them. This is left for the body of the study.

The Nature of the Study

There can be no doubt that some technologies present hazards. But conflicts about these technologies and their hazards cannot be treated merely as technical disagreements about the acceptability of those hazards. Otway & von Winterfeldt (1982, p. 255) put it this way:

> The acceptable risk formulation has provided increasingly elaborate and precise answers to the wrong question; future research on the acceptability of technologies must be linked to the critical questions of institutions, participation and policy implementation, for until sources of conflict have been addressed, the decision taken--no matter how technically rational--may ultimately be only empty prescriptions lacking arrangements for their realization.

The trouble with basing analysis on questions of "institutions, participation and policy implementation" is that there is no widely accepted theory of technological development or decision making about technological hazards. (Jungermann, von Winterfeldt & Coppock, 1982, p.13) Nevertheless, more than a decade of experience with technology assessment and risk analysis has accumulated. It ought to be possible, therefore, at least to order the various kinds of information required for rational decisions about technological hazards, and perhaps to describe appropriate analytical approaches for gathering and interpreting these different kinds of information. It is such an ordering that is referred to in this study as a "conceptual framework".

The style, or methodological approach, of this study is determined by its subjects. Although the study addresses the management of technoligical hazards, it is based neither on specific technologies nor specified hazards. Rather it addresses the generic treatment of technological hazards, the structuring of arguments and data rather than the drawing of conslusions from specific data about a given case. Since it focusses on the "meta-question" of the management of technological hazards in general, it is necessarily analytical rather than empirical in nature, even though empirical data and actual experience are used to illustrate various points.

"Empirical" research refers to an approach whose conclusions are based on data collected either through experimentation or observation of naturaly occurring phenomena. Empirical research usually postulates a hypothesis, or explanation of the phenomenon under examination, and then tests this hypothesis by seeing whether the data reveal conditions corresponding to the hypothesized explanation. This is the classical approach of the natural sciences. Questions of the relationship between data collection and explanation, and the reconciliation of alternative explanations of the same phenomenon, are relevant but difficult. Such issues will be mentioned here only when they pose unavoidable questions concerning procedures for the collection or interpretation of data relevant to decisions about response to technological hazards.

Although it will be assumed that rational and intentional response to technological hazards requires empirical analysis, this study itself is not empirical. Rather, it is an example of "analytical" analysis. The conclusions of analytical studies need not be based on the results of data collected (although they may). The "test" of an analytical result may be simply the plausibility of its description of a phenomenon previously only poorly understood. Analytical results are usually tested against empirical data whenever possible. It is possible to produce analytical results for phenomena for which no data exist, and for which it is not clear how empirical data might be collected. For example, some natural science research attempts to explain the origin of the universe. While it is possible to compare certain phenomena we can observe with the logical consequences of hypothesized beginnings (e.g., the "big bang" theory or that of cyclical expansion and contraction), it is not possible to observe the actual event itself.

When research or analysis is applied to individual or collective human processes, it is possible to distinguish between descriptive and prescriptive research. A "descriptive" approach attempts to describe how the the process being examined actually occurs. A "prescriptive" approach attempts to describe the best way to organize and conduct that process, or at least an improvement over what occurs by happenstance. One danger of the prescriptive approach is that procedures can be devised which have no relation to what the people involved are willing to undertake. This is generally how "empty prescriptions lacking arrangements for their realization" come about. Thus, although this study is essentially prescriptive in nature, attempting to improve the management of technological hazards, it retains a strong descriptive flavor. In this way it is hoped that the suggestions proposed are within the realm of the possible.

This study attempts to describe how rational decisions about response to technological hazards ought to be made, not necessarily how decisions about such hazards actually happen. Thus, a demonstration that the approach here has never been used should not be considered a refutation of these ideas. However, the study is also based on the assumption that if its conclusions are to be realistic and useful, the analysis may not depart too far from those methods which are in current use. In practice, this means that wherever possible each major concept is illustrated with examples drawn from empirical materials.

It should be noted that from time to time the word "rational' or the combination "rational and intentional" will be used to describe the study or some part of it. There are myriad definitions of rationality, and the use of these terms should be explained. The most fundamental concept in the study is the idea of personal volition, of conscious decision on the part of the individual. Such a decision is rational and intentional if the individual attempts to understand the consequences of implementing his decision. The discussion up to now has emphasized the complexity of technological hazards. Not only must a variety of different kinds of information be incorporated, but political motivations of a variety of actors functioning at different levels should also be considered. The conceptual framework developed here provides a guide not only to the different kinds of information required, but also to the shifting of analytical focus from such individual choice through various levels of aggregation so as to construct an image of potential actions at different levels of society.

Although the fundamental basis for decision making in this study is personal volition, the decision process will normally be treated in a "top down" manner. That is, human aggregates will first be examined an the most general level relevant to the issue in question. After this, each social entity will be broken down into its components if greater detail is necessary, until individual decision makers are ultimately reached. For example, the relevant actors influencing response to oil pollution on the high seas include nation states, oil companies, tanker owners, ship masters and seamen. In practice, it would probably prove necessary to examine some of these groupings in greater detail, for example, the change over time in the ministries or public agencies representing key nation states. The assumption is that all decisions are taken by identifiable individuals, functioning in a specific social setting, no matter the level of aggregation at which the decision is to be implemented. Personal volition thus refers to the "public person" interacting in a specific situation and necessarily results in some observable consequence (including inaction). Individual reaction, on the other hand is a "private" phenomenon which need not affect behavior.

An important attribute of this study is the conceptual level at which the anaysis focusses. Over a dozen different analytical approaches are referred to in the following chapters, and as many additional concepts presented which are not surrounded by the assumptions, methods and styles of analysis which make up an analytical approach. For several of these research approaches (for example, cost-benefit analysis or psychometric surveying), the number of books and articles dealing with various aspects of the approach is rather large. This study makes no attempt to present a comprehensive review of the professional literature of each approach. Rather the aim is to present as clearly as possible the major elements of an approach, so that its contribution to the overall aim of managing technological hazards might be seen in relationship to other approaches. The central question of the study involves the interpretation of diverse information from a variety of sources, collected and manipulated according to the methods of several different approaches. The scientific discussion of those methods, while by no means irrelevant to an overall evaluation of the application of the approach, often contributes little to its integration with different approaches. For example, there are several alternative methods of determining discount rates in cost-benefit analysis, and the choice of one can have an important influence over the outcome of the analysis. However, the detailed examination of their respective advantages and disadvantages under different conditions does not shed much light on the comparison of the results of a cost-benefit analysis with an engineering feasibility study or a public opinion survey.

The Structure of the Presentation

The following chapters correspond roughly to the development of concerns in the area of risk analysis. It has already been mentioned that this began with a focus on the technical issues involved, shifted to the psychological reaction to the technologies and their hazards, and most recently turned to the associated societal or political conflicts. This is the general order in which the relevant topics are addressed in the subsequent chapters.

The remaining chapters could be broken into four parts. The first (Chapters 2 and 3) reviews the development of risk analysis and the general analytical underpinnings of the management of technological hazards. It thus looks at the approaches which have been applied, together with their limitations and weaknesses. The second part (Chapter 4) presents a conceptual framework intended to enable the various relevant elements to be placed in relationship to one another. The proposed framework is based on typical interactions in technical conflicts, decision making in a pluralistic setting, and an ordering of the topics relevant to policy and operational decisions about technologies and their hazards. Then in the third part (Chapters 5, 6 and 7), concepts and approaches are presented that enable the application of the conceptual framework in a manner which goes beyond the analysis commonly found in current approaches. These are discussed at the level of individual reaction, collective response, and systemic influences. Finally, the concluding chapter presents a general summary and draws conclusions. Again, the overall aim is to develop a conceptual framework in which both scientific and technical judgments and political judgments can be seen in perspective.

It should be mentioned that an alternative order of presentation might, in certain respects, be more advantageous. The overall thrust of the study is that the management of technological hazards must be seen as a political as well as a technical process. Thus, the proper context for the analysis of technological hazards may be said to be the political judgment required. However, recognition of this has come quite late in the development of the field. By following the rough chronological order of developments, this overall context comes last. For this reason, the political context has been briefly introduced in this chapter. However, a full discussion is left to Chapter 7.

Chapter 2, "Technological Hazards and Their Evaluation", reviews the development of the field of risk analysis. It begins with the basic concept upon which risk analysis is based: that one can intervene in the sequence of cause and effect to reduce the likelihood of technological hazards or to mitigate their consequences. This reflects an emphasis on the examination of technical options. The next topic addressed is the evaluation of alternatives, and introduces the idea of comparing the various advantages and disadvantages in some way. The next topic is the determination of acceptability, and goes into both the guides for judgments used in the courts and other adjudicatory processes as well as into psychological influences on acceptability. Next, a dynamic element is introduced: the idea that it is inadequate to think in terms of a single decision, but rather that technological hazards must be managed over time. The final topic in Chapter 2 is the interaction of the management of technological hazards with the political process.

Chapter 3, "Formal Analysis of Technological Hazards", describes the analytical approaches that have been commonly used in examining technological hazards. The chapter might be divided into two sections, the first dealing with conceptual bases for analysis and the second focussing on the organizational perspective of the analyst. Two conceptual approaches are described: safety and reliability engineering, and cost-benefit analysis. Following this, two perspectives are described: that of society as a whole, and of an individual enterprise. Finally, a few observations are made concerning the limitations of formal analysis.

Chapter 4, "An Analytical Framework for Decisions Concerning Technological Hazards", suggests a structure within which both the scientific and technical judgments required in the analysis of technological hazards and the political judgments required in the determination of proper response to them can be seen in relation to one another. The chapter begins with a simple model of the typical patterns of interaction among relevant groups in technological issues, and then presents a set of guidelines for making decisions under such conditions. Based on this, a framework is presented which incorporates the various factors that are likely to play a role the analysis and management of technological hazards.

Thus, Chapters 2, 3 and 4 address the first aim of the study: to develop a framework within which both the technical or scientific judgments and the political judgments required can be given proper emphasis. Chapters 5, 6 and 7 address the second aim: an exploration of the factors important in the political judgments applicable to the management of technological hazards.

Chapter 5, "Individual Reaction to Technologies and Their

Hazards", reviews several lines of research relevant to the description of individual reaction. The first of these is based on psychological studies of judgment, and describes both the cognitive attributes of decisions about risk and the typical differences found among different groups of individuals. The second set of research findings deals with attitudes and the ways behavior relates to them. The final body of research described in Chapter 5 deals with values and value change on a fairly wide basis in society, describing behavioral consequences one might expect should the postulated value change have occurred.

Chapter 6, "Collective Response to Technological Issues", moves to the level of group response. It is argued that individual action will seldom be a sufficient response to technological hazards, so that only collective response is delt with in the study. Chapter 6 describes the various kinds of selective incentives that can motivate collective response with respect to technological hazards. It describes the basic parameters of choice, the possible interdependencies, the potential strategies, and the alternatives ways of coordinating collective action.

Chapter 7 describes the "Systemic Attributes of Modern Technological Conflicts" in terms of the basic underpinnings of ecocentric and technocentric views of technology, of the politicization of science and technology, and of the setting of social conflict in modern industrial societies. Several of the ideas presented in detail in Chapter 7 have already been mentioned briefly in this chapter. Finally, Chapter 7 discusses the influence of personal experience on perception and on preference for specific strategies.

Finally, Chapter 8, "Orientations and Outlooks", presents a

general summary of the various factors relevant for the managment of technological hazards and of the conditions under which it may be expected to be conducted.

2 TECHNOLOGICAL HAZARDS AND THEIR EVALUATION

Throughout human history individuals and their representatives in various kinds of groups--in primitive tribes and modern governments alike--have examined specific hazards, evaluated the risks associated with them, and decided how to respond. However, as complex technological structures and activities have come to dominate our lives, we have naturally applied the same logic to dealing with the problems that we used in the original development of the technologies themselves. Thus, we commonly respond to technological hazards with rational analysis of the causes and effects and how one might best respond to them.

It could be argued that no methodological framework can be comprehensive enough to apply to all actual situations which may be said to involve technological hazards. Nevertheless, exactly because of the complexity of the topic and the necessity of incorporating a variety of information from different sources and analytical approaches, a body of literature has emerged dealing with the analysis of technological hazards and responses to them. This chapter presents a critical overview of the key ideas in that literature, attempting to describe the concepts which provide the basis for rational assessment of technological hazards and the possible ways of dealing with them as well as some of the weaknesses of these concepts.

It should be noted that the literature referred to in this chapter is limited to that dealing with the development of the approach of risk analysis or assessment. The attempt to include everything relevant to technological hazards would of course mean referencing a virtually unlimited number of sources. Therefore it is necessary to be more restrictive, and this chapter focusses on the approaches proposed for the evaluation of technological hazards and responses to them. This corresponds with the overall aim of the study: to encourage rational and intentional treatment of technological hazards. Later chapters will refer to broader sets of literature in so far as they are relevant for the specific issues raised by the management of technological hazards.

Intervention in Chains of Cause and Effect

Certainly the most fundamental idea of analysis is that initiating events cause other things to happen. The basic task of analysis is to take a more or less complex situation and to reveal the elements and their interactions. The method of analysis is to produce a simplified model of the chains of cause and effect in the phenomenon under examination. Analysis usually begins with the examination of one link in such a chain of cause and effect, and through discovery of the relevant interactions of various elements, proceeds to build up a picture of the overall chain. In part, this is a pragmatic consequence of the way technological hazards are generally discovered. Some are recognized initially as known or suspected dangers (such as carcinogenic substances) and the analytic problem is to discover where and how their effects are distributed. Others become known because their effects become apparant (such as local, abnormally high incidence of disease) and the analytic problem is to determine the cause and source of these effects.

The analysis of hazards and consequences can be described in general terms as the collection of information about four components of technological hazards:

> The nature and extent of the <u>source</u> of danger or potential damage
>
> The chain of events, <u>pathways</u> and processes that connect the cause(s) to the effect(s)
>
> The relationship between the characteristics of the impact (e.g., dose) and the <u>effect</u> (e.g., response)
>
> The nature and strength of individual and group perception of and <u>reaction</u> to the overall sequence from source to effect

Subsequent chapters will describe a variety of methods that are used to produce information about sources, pathways, effects and reactions as well as ways that such information can be evaluated and used in designing appropriate responses to technological hazards. But this elementary characterization suffices for the purposes of setting out the basic concepts necessary for analysis.

Following logically from the sequence of source, pathway, effect and reaction is the concept of intervention. Figure 2.1 presents a simple characterization of hazard causation and response which attempts to incorporate a broad range of factors which influence intervention measures. These factors range from the human wants which motivate the selection and use of a technology to the ultimate consequences of exposure to the hazards associated with it.

The model of hazard causation and response in Figure 2.1 encourages looking "upstream" to the possiblity of altering the

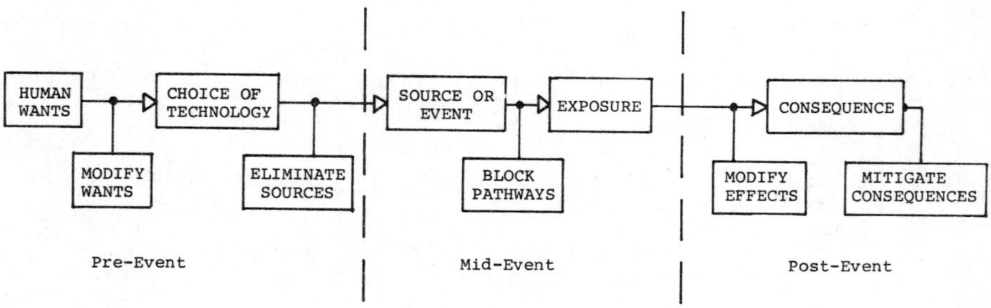

Figure 2.1 A Model of Hazard Causation and Response
Adapted from: Fischhoff et al, 1978

motivation for using the technology in the first place as well as "downstream" to ways of blocking the effects of the hazard or of mitigating its consequences.

Intervention programs can be further distinguished according to: 1) the point of application in the sequence in Figure 2.1, 2) the temporal variation in control, 3) the spatial variation of control, and 4) the instruments of control. (See Grad et.al. 1975, Chapter 1 for a general discussion of these dimensions.)

The major point here is that there are usually a variety of possible responses to a given hazard, ranging from preventing its occurrece in a varitey of ways to limiting its effects and repairing the damages or otherwise mitigating the consequences.

Evaluation of Alternatives

The fact that, at least conceptually, there are a whole range of possible responses to a particular technological hazard introduces the question of how to choose among them. Furthermore, the characteristics of the sequence of events from source to consequence and the variety of possible interventions mentioned in the previous section suggest that this choice can also involve the basic decision of whether to attempt to avoid exposure to the hazard altogether or to cope with its consequences. Thus, the evaluation of alternative responses to technological hazards requires the balancing of social as well as technical considerations. One expression of this has been the attempt to compare the social costs and benefits of technologies, their hazards, and possible intervention programs.

In 1969, Starr published an article in Science proposing a method for extending the analysis of technological developments from the relationship between expected performance and monetary costs to the inclusion of all social costs. He attempted to establish a quantitatve measure of benefit relative to cost for accidental death due to technological developments in public use. To accomplish this he compared accident data for a variety of technologies with the expeditures devoted to safety measures. His conclusion was that "this methodology for revealing existing social preferences and values may be a means of providing the insight on social benefit relative to cost that is so necessary for judicious national decisions on new technological developments." (Starr, 1969, p. 1237)

However, the "revealed preference" method proposed by

Starr has been criticized on several points. (Otway, 1980) The method assumes that society has somehow arrived at an optimal balance of risk and benefit through some iterative process of negotiation and exchange. Therefore, a proper balance for new technological hazards can be determined by calculating the existing trade-offs for similar technologies. However, this assumption overlooks what appears to be a shift in public opinion and preference concerning technologies and their attendant hazards. Furthermore, the existing trade-offs have for the most part resulted from political processes which likely reflect the preferences of established institutions and decision makers rather than the public at large. Thus, if the public is indeed now more inclined to exert its own will than was previouslly the case, reliance on "revealed preferences" would likely meet with considerable public opposition.

One response has been to be more analytical in the balancing of benefts and risks. In 1971, the Committee on Public Engineering Policy of the American National Academy of Engineering conducted a colloquium designed to examine the integration of cost-benefit approaches into decision making about technological hazards. That meeting set the general framework of developments in risk assessment in the following terms (Nat. Acad. of Eng. 1972, pp. 3-4):

> Technology Assessment: the systematic determintion of the consequences of various alternative actions involving the application of technology
>
> Benefit-Cost Analysis: the determination of <u>all</u> the benefits and costs of proposed actions (thereby being broader than traditional cost-benefit analysis involving only economic factors)

> Benefit-Risk Analysis: that category of benefit-cost analysis for which risks to health are important components of the costs

This notion that the tradeoffs to be made should include special emphasis on risks to life and health remains an important characteristic of risk assessment efforts today.

Perhaps the most complete attempt to present a methodological framework for the analytical evaluation of technological risks is to be found in An Anatomy of Risk . (Rowe, 1977) The basic premise of this approach is that the agencies of government responsible for setting standards of acceptability for risks of various kinds must make three kinds of judgments: technical judgments (required by incomplete or inconclusive technical and scientific information), societal judgments (required to balance benefits, costs, and the like) and managerial judgments (required to meet the exigencies of implementation and enforcement). Rowe (1977, p. 366) describes a number of techniques for making such judgments, and proposes a four-step approach to risk evaluation:

> 1. Direct Gain-Loss Analysis: made by the sponsor of the technology; involves economic cost-benefit analysis of voluntary exposure to risks
>
> 2. Indirect Gain-Loss Analysis: made by the government office proposing action; attempts to accomplish overall societal assessment; often qualitative, achieving consensus about four levels: favorable (balance overwhelmingly favors benefits over costs), marginal (balance only slightly positive or even), unfavorable (costs generally outweigh benefits), unacceptable (costs far outweigh benefits)
>
> 3. Cost Effectiveness of Risk Reduction: assessment of the degree to which risks can be reduced at specified levels of expense and effort
>
> 4. Reconciliation of Risk Inequities: if those who assume the risks are not those receiving

> the benefits, the risk should be identified and its nature and type ascertained and compared against the levels of risk of that type extant in society (i.e., a risk referent)

Rowe observes that the first three steps in the above approach are well described by current practice, but that there are two basic kinds of judgments involved throughout: those contributing to the estimation of the hazard and its magnitude, and those contributing to the evaluation of the degree to which exposure to that hazard is undesirable.

Determining Acceptability

Two general approaches have been identified for determining the acceptability of particular levels of exposure to technological hazards. (Otway, 1980) The first, the revealed preference method, has already been mentioned above. The second is called the "expressed preference" method, and uses survey tehcniques to obtain directly from people their preferences concerning the hazards in question. Just as the revealed preference method presumes that previous decisions about existing hazards were rational and well-informed judgments about the acceptability of the risks involved (an assumption subject to reasonable doubt), the expressed preference method is also based on questionable assumptions. Namely, it presumes that the respondents already have an opinion about the hazard and its consequences, and that they are able to express that opinion clearly and without confusion. While this may be the case for some technological hazards, it is quite unlikely for new technologies, especially those departing radically from current practice, or for consequences with indirect and confusing chains of cause and effect.

In part because of these difficulties in judging acceptability, the Committee on Science and Public Policy of the National Academy of Sciences in the United States undertook in 1973 a project to examine the notion of safety and how safety determinations are made. One of the results of this effort was the identification of several "guides to acceptability" which are used in various kinds of judgments. (Lowrance, 1976, pp. 76-86) These are listed below.

Reasonableness -- the frequenly invoked "rule cf reason" is in a sense a phantom citation: it provides the decision maker with little guidance as to what is reasonable and what is not

Custom of Usage -- an example is the "generally recognized as safe" (GRAS) list of food additives which are considered safe simply because they have been used with no recognized negative effects

Prevailing Professional Practice -- long established as the criterion for physician's clinical practice, this idea is also applied in matters of engineering and manufacturing design

Best Available Practice -- also called "highest practicable protection" or "lowest practicable exposure"; this standard has often been applied with respect to emissions controls which should employ "best available means"

Degree of Necessity or Benefit -- applied with respect to food additives, this criterion specifies that no additives should be used unless its benefit has been demonstrated

The Delaney Principle -- part of an ammendment to the U.S. Food and Drug Act, this requires that "no (food) additive shall be deemed to be safe if it is found ... after tests which are appropriate for the evaluation of the safety of food additives, to induce cancer in man or animal"

No Detectable Adverse Effect -- many hazards, such as moderate levels of exposure to X-rays, were previously said to have "no detectable adverse effect", a very weak guideline which essentially recognizes the lack of knowledge

Toxicologically Insignificant Levels -- of similar dubious nature to the criterion of "no detectable adverse effect"

The Threshold Principle -- if it can be proven that there is a level of exposure below which no adverse effect occurs, then subthreshold levels of exposure might be considered safe; the problem lies with demonstrating the threshold

Exposure Relative to Natural Background -- deriving from radiation standards, this claims that if exposure is kept below the natural background level, it may be assumed that additional consequences will neither differ in kind nor exceed those encountered throughout history

Occupational Exposure Precedent -- occupational records are often used to estimate levels of exposure that can be borne without ill effect

Public Referenda and Polling -- the direct use of the expressed preference method, this has been used in actual determinations of safety standards in very few instances

Comparison with Accustomed Hazards -- the revealed preference method; has been proposed but not used as the formal basis of safety standards, except in the loose sense in which several of the above criteria might be said to have this as their aim

In addition, the National Academy of Sciences study identified several kinds of considerations that influence the way individuals react to hazards in performing safety judgments. Figure 2.2 presents the array of considerations developed in the project.

The recognition that individuals react differently, depending on whether risk is assumed voluntarily or involuntarily, whether its effect is immediate or delayed, and so on has led to considerable research aimed at identifying the determinants of psychological reaction to exposure to various kinds of hazard.

A recent study, which is typical of one line of research on the determinants of reaction to risk, examines 90 technological and other hazards in terms of eighteen different characteristics. (Slovic, Fischhoff & Lichtenstein, 1980) The risk associated with each hazard, the eighteen risk characteristics, and the benefits that might be expected from the

Risk assumed voluntarily	______	Risk borne involuntarily
Effect immediate	______	Effect delayed
No alternatives available	______	Many alternatives available
Risk known with certainty	______	Risk not known
Exposure is an essential	______	Exposure is a luxury
Encountered occupationally	______	Encountered non-occupationally
Common hazard	______	"Dread" hazard
Affects average people	______	Affects especially sensitive people
Will be used as intended	______	Likely to be misused
Consequences reversible	______	Consequences irreversible

Figure 2.2 An Array of Considerations Influencing Safety Judgments

Source: Lowrance, 1976

activity associated with each hazardous technology were rated. A factor analysis of the ratings reveals a grouping into three categories. The first is a group of twelve which suggests that a hazard whose risks are believed to be difficult to control are also seen as dreaded, catastrophic, hard to prevent, fatal, inequitable, threatening to future generations, not easily reduced, increasing in frequency of occurrence, involuntary, and personally threatening. The researchers label this group of factors "dread". A second group of five characteristics which relate to each other more closely than to the other factors is: observability, knowledge about the hazard, immediacy of consequence, and familiarity. This group is labeled "familiarity". Finally, a single characteristic stands alone: the number of people exposed. The researchers call this the "catastrophic potential". Slovic and his colleagues conclude that these three factors--dread, familiarity, and catastrophic potential--are the pftg important influences on the psychological reaction to technological hazards.

Other research has employed the attitude-behavior model to demonstrate that strongly pro- and anti-nuclear power groups base their respective opinions on quite different considerations. (Otway, 1980) The findings reveal that individuals in favor of nuclear development tend to emphasize technological and economic benefits, and environmental and physical risks in their judgments. In contrast, those opposed stressed psychological aspects of the problem and the socio-political implications. This suggests that, even though the nuclear debate appears to focus on a discussion of a specific set of hazards, arguments based on rigorous safety standards might be totally misplaced because the reactions of those opposing nuclear development seem to be based on entirely different concerns.

Ongoing Management rather than Single and Separate Choices

Many of the analytical tools which are used to evaluate technological hazards have been developed principally for aiding investment decisions. (See Chapter 3 for a description of the most important of these.) As such, most are oriented toward a single assessment of the advantages and disadvantages of a proposed project. Few are designed with an ongoing series of decisions in mind. However, it appears that management techniques rather than investment analysis are more appropriate for the major problems surrounding technological hazards.

In the not too distant past, decisions about the proper response to all sorts of hazards were left to technical experts. Today, such expert decisions are increasingly being challenged. In fact, it appears that at least part of the dissatisfaction on the part of the general public is with the procedures by which such decisions are made rather that with the hazard per se. (Wynn, 1980, p. 196) Thus, securing a consensus seems to be an important part of the process of making and implementing such decisions. However, everything suggests that debates about technological hazards will develop divergent, polarized positions rather than move towards convergent conclusions. (Pearce, 1980, p. 63)

One result has been the attempt to systematically examine such conflicts and their consequences for decision making structures. For example, Nelkin & Pollak (1980) have examined a number of procedures which have been tried in various places with the aim of securing consensus about technological hazards. Their conclusion is that it is difficult to generalize about such procedures themselves, but that there appear to be certain conditions which allow dissenting groups to communicate effectively with administrative agencies. These include a careful definition of the problem giving due weight to political considerations, a fair distribution of expertise, and a real margin for choice. (Nelkin & Pollak, 1980, p. 75) Although these conditions do not stimulate a consensus by themselves, they do appear to provide at least an atmosphere within which constructive discussion can proceed.

O'Riordan (1979) has described the functions of risk management as identification, estimation, evaluation, and control. His characterization of the interrelationships among these functions may be found in Figure 2.3. The key feature, according to O'Riordan, is the incorporation of both scientific analysis and political judgment.

The most logical place to begin the examination of the circle of risk management functions in Figure 2.3 is at the point of risk identification. This important activity is obvioulsy dependent upon knowledge, research findings, experience and

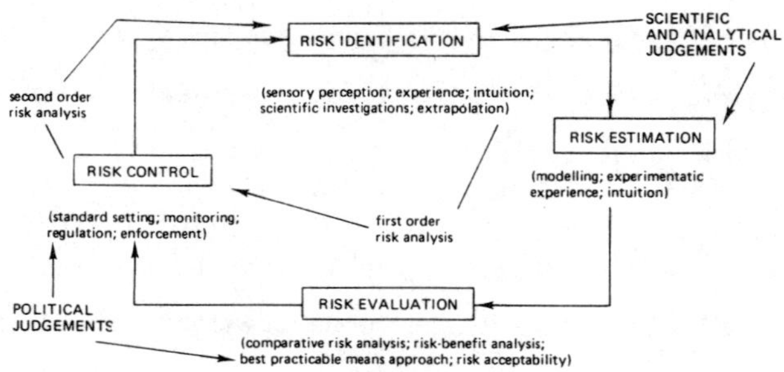

Figure 2.3 Risk Management Functions

Source: O'Riordan, 1979, p. 23

intuition. But it is unlikely that even the most sophisticated studies will identify all the relevant hazards, so it is important to include the search for "second order risks"--the risks associated with the attempt to reduce the immediate or first order risks.

Risk estimation involves the description of of risk causes and pathways, the likely consequences, and their distribution in space and time. As has already been indicated, this can be extremely complicated and require considerable time and effort.

Risk evaluation involves not only the social judgment of risks as identified and estimated, but the balancing of those risks against expected social gains. A number of procedures have been proposed for this task, some of which will be described in later chapters.

Risk control is the phase of implementation where regulation, monitoring, enforcement, and re-evaluation take place. Ideally, these tasks are undertaken after the research, consultation, and judgments of the other phases have been completed. But in many instances the process is short-circuited due to lack of time, political expedience, insufficient funds, or adherence to established procedures regardless of their suitability.

Although this formulation of the functions of risk management--as well as most of the other concepts and approaches presented up to this point--has been in terms of the assessment of overall societal advantages and disadvantages, there has been a parallel development at the level of the individual firm. Within the self-interested motivation of the firm, the concept of risk and insurance management aims at reducing or eliminating financial losses due to accident, damage, etc. In more precise terms, risk and insurance management deals with "pure" risks (events which can lead only to losses; for example, fire, accident, theft, neglegence) as opposed to "speculative" risks (events which offer both the chance of gain and loss, such as change in technology, in consumer taste, or in government regulation). (Head, 1982, pp. 181-182) Olson & Simkiss (1982, pp. 117-118) characterize the procedures of risk and insurance management as follows: 1) identification of the pure risk events that may pose financial losses on the enterprise; 2) evaluation of the potential costs that would be involved if such events should occur; 3) estimation of the frequency with which such events might occur; 4) analysis of the alternatives that are available; 5) selection of the most desirable alternative in light of the overall goals of the enterprise at the time the decision is made; and 6) periodic re-evaluation and monitoring. The similarities between this "private sector" approach and the "public sector" studies described above are obvious. The self-interest of the individual enterprise limits its attention to a rather more restricted set of consequences than is commonly the case for public assessments, and the strategies of response are similarly restricted. But in generic terms, there is little difference in the problem addressed in the public and private approaches: what is the appropriate response to the hazard under question.

Risk Management and the Political Process

Although the thrust of "public" and "private" assessments may be the same, they differ dramatically in one respect. In private assessments, the legal framework and general political values are usually taken to be outside the realm of influence of the firm, and such factors are treated as exogenous variables, beyond the influence of the study. Public assessments, however, often strive to incorporate such factors into the variables evaluated in the examination. In fact, the more that individual and group reaction to hazards become important, the more likely it will be that some degree of public consensus will be required concerning programs to correct or ameliorate the situation. Of course, this brings the process into the realm of politics, or alternatively, politics into the analysis. The fact that this appears to be the case for so many environmental hazards is what brings O'Riordan to claim the combination of scientific analysis with political judgment to be the most distinctive characteristic of risk management. (O'Riordan, 1979, p. 260)

It seems clear that the most preliminary step in the identification of hazards, the recognition of general areas which should be examined, is subject to considerable "input" from the political arena. The environmental movement of the 1960s and 1970s has largely influenced both public and private institutions in their treatment of broad classes of technological hazards. (Bowman, 1976; O'Riordan, 1981; Mayer-Tasch, 1976) Similarly, the recall of defective products by individual firms follows closely the treatment of strict liability by the courts, which in turn appears closely related to the success of the consumerism movements in various countries. (Damary & Hurst, 1982)

The functioning of what has here been called the political process in identifying general areas of potential hazard appears, however, to be somwhat accidental and perhaps arbitrary. The simplest way of characterizing environmental and other special interests for this study is to draw an analogy to an economic cartel. Special interests may be said to "purchase" legislation in exchange for support at the polling place. In the American system of presidential democracy, this "market" is very strong because each legislator runs on his own position and is somewhat independent of the party platform (as to some extent, is even the presidential candidate), Party discipline is relatively weak. In parliamentary systems, on the other hand, party discipline is strong because of the necessity of holding a majority coalition together in order to form the government. In parliamentary systems, it is the party rather than the individual legislator that functions as the "seller" of legislation. In either case the "market" for legislation is subject to acute free-rider problems. A citizen, group or firm which receives advantage from a particular piece of legislation usually receives that advantage whether or not any express contribution was made to its enactment. The situation is similar to the economic conditions which encourage development of a cartel. This analogy describes both the establishment of a political coalition within the legislature on a particular subject (necessary for enactment of special interest legislation) and for the negotiation of the "price" of legislation to those outside the legislature. The analogy is particularly appealing because the results of much special interest legislation are quite similar to what would be achieved by a marketplace cartel. (Posner, 1977, p.406)

Economic theory of cartels would suggest that special interest groups within the environmental, consumerism and other movements would have different degrees of success. One would expect, for example, that groups opposing the siting of a particular facility, say a power plant or a motorway, would be more effective than would consumer interest groups. Indeed, the inadequacies of of public consumer policy is often lamented. (Matthoeffer, 1977; Biervert et.al. 1978) A successful economic cartel has a limited number of participants. Too many participants are too hard to discipline, and it is difficult to keep track of individual changes in price or quality and the consequent changes in relative advantage with respect to the other cartel members. Perhaps the best-known cartel today is the Oil Producing and Exporting Countries (OPEC). Even with the obvious power of OPEC to influence, almost unilaterally determine, the price of crude oil, the cartel has not been able to maintain a single price for a barrel of crude among its members.

But "consumers" as a group are too numerous, with far too diverse interests, to form an effective interest group. It is relatively easy to recognize the other individuals who will be affected adversly by the construction of a particular facility. But it is much less easy to recognize the other individuals affected by a defective consumer product. Legislative or administrative actions dealing with neglegence or defect associated with a particular product would probably be of interest only to the consumers of that product. The next action, dealing with a different product, would likely interest an entirely different set of consumers. Thus, interest groups for consumer protection are usually made up of small groups with interest in the concept of consumer protection rather than in particular products. In looking at consumer interests overall, there are neither small, disciplined groups of consumers with an adequate influence on the demand for a particular product, nor large clearly defined groups which have the ability to define their membership and protect their advantages. What has been accomplished in the area of consumer policy has mostly been the result of expert and professional groups motivated to improve what they see as inadequacies in the system of production, distribution, and consumption rather than of groups of disgruntled consumers who have banded together. The lack of a clear link between such policy goals and the interests of consumers is recognized as a major weakness of what has been accomplished. (Rock et.al. 1980)

Despite the differences in effectiveness of interest groups of various kinds, they will continue to identify areas for examination in the identification of technological hazards. Thus, at the beginning and the end of the process of risk management, at the stages of identification and of implementation, we can observe the strongest interaction between analysis of hazards and political processes. (This is because the other primary point of interaction, evaluation, is often one in which surrogates are used in place of true interaction with citizens.)

Summary and Conclusions

This chapter reviews the principal ideas of risk analysis as they have appeared in the literature. But since not all of these ideas receive equal emphasis in this study, those considered most important should be briefly summarized.

The most fundamental concept of risk analysis, and a basic assumption in this study, is that an accurate description of technological hazards and responses to them require reference to all of the following: the source events, the pathways of cause-and-effect resulting in exposure, the consequences of exposure, and the reactions to each of these on the part of affected persons. Information is needed about each of these for a balanced assessment of a hazard and response to it.

A second basic position of this study draws upon the distinction between "pure" and "speculative" risk as defined in risk and insurance management. Although the perspective of this study is that of society rather than of private enterprise, greater emphasis is still given to "loss producing events". Of course, "loss" often means something quite different for society as a whole than for an individual firm, but the emphasis remains on the negative and undesirable consequences of events and conditions. In particular, the focus is on damage, injury, illness, and and early death.

Another important distinction is that among the various functions of risk analysis: identification, estimation, evaluation and control or implementation. Each of these functions requires somewhat different information, employs distinctive analytical methods, and assigns slightly different importance to the various relevant factors. One of the most important characteristics of risk analysis as a field of endeavor has been the differential development of these functions as separate topics. For example, the literature reveals considerable consensus about methods for the identification of hazards and the estimation of their likelihood. But it also shows a large difference of opinion about how the consequences should be evaluated and control programs organized. This has sometimes been characterized as a paradox: at the same time that the degree of agreement about the extent of exposure and its potential consequences has increased dramatically, the agreement about the acceptable level of exposure has decreased markedly.

Part of this disagreement, at least in terms of methods for evaluation of the consequences, has been the debate about the use of the revealed preference approach as opposed to that of expressed preference. In some respects the revealed preference approach is simpler. It enables a technical comparison between two or more sets of hazards, a task that can be performed with all the logic and precision that engineering offers. But in at least some cases, other considerations override such technical evaluations. Perhaps for those technological hazards that never become public issues the revealed preference approach might suffice. But when conflict emerges, reliance on technical factors alone will provide an inadequate characterization of the choice.

Research has revealed that people's reaction to technologies and the hazards associated with them are complex. Nevertheless, there appear to be certain sets of factors which have greater influence than others. In particular, the characteristics of "dread", "familiarity" and "catastrophic potential" appear to play an important role in people's reaction to technologies and their hazards.

Finally, it should be noted that public discussion about technological hazards and response to them has gone through three phases. The first emphasized mostly technical aspects of hazards and response to them. The second devoted more attention to the evaluation of exposure to these hazards, especially in comparison to other hazards. Most recently, greater emphasis has been given to the conflicts associated with such hazards, including programs of response to them. Thus, one can say that there has been a trend from the technical to the political. This study corresponds to that development. It presents an analytical framework intended to incorporate both the technical and scientific judgments on the one hand and the political judments on the other, as well as examining in greater detail the factors influencing relevant political judgments.

3 FORMAL ANALYSIS OF TECHNOLOGICAL HAZARDS

Systematic analysis of hazards is not new. This chapter reviews the most common approaches to the analysis of technolgical hazards and the decisions about how to respond to them.

The Scientific Basis of Health and Safety Regulation (Crandall & Lave, 1981) is a collection of papers dealing with the use (or abuse) of technical information by policy makers in the United States. It examines a series of cases from three separate perspectives. For each case, the scientist's view is presented first, then the economist's, and finally, that of the regulator. This order implies a model of the policy making process that is widely shared. Namely, that scientific research, sometimes independent of any policy motivation, reveals the existence of some health or safety hazard. The resulting scientific data and conclusions become the basis for an economic analysis of costs and benefits, and both are used by regulators as bases for their actions. (Hadden, 1982) This implicit model of the scientific and economic underpinnings of health and safety regulations is similar to the view of policies concerning technological hazards in general which is used in this study.

However, as is true of any field of systematic analysis, the more a topic is differentiated, the more difficult it becomes to define universal generalizations. So it should not be surprising that a bit more than two decades of analysis devoted to technological hazards and responses to them has led to a set of concepts and techniques rather more complicated than the simple characterization above might suggest.

This chapter reviews the principal approaches to the formal analysis of technological hazards. The first section addressed the methods commonly used to organize information about the technical attributes of such hazards. The second section then turns to the economic evaluation of actions and consequences. In the third section, approaches which can compare alternative action programs from the perspective of society as a whole are examined. This is followed in the fourth section by an examination of the evaluation of response programs from the perspective of the enterprise. Finally, in the last section, certain conceptual limitations to formal analysis per se are examined.

Safety and Reliability Engineering

In the 19th century, the safety of an employee was generally considered to be his own responsibility. This was supported by the rules of compensation that were prevalent in common law doctrine of the time. The "fellow-servant" doctrine, for example, excused the employer from responsibility for injuries resulting from the action of a fellow worker. The "assumption of risk" doctrine held that in accepting employment the worker assumed any risks associated with it. Finally, the "contributory negligence" doctrine absolved the employer from responsibility for injuries resulting from accidents whenever the employee had contributed to the cause of the accident. (Encyclopaedia Britannica, 1981, p. 137)

Social consciousness changed, however, and by the early 20th century workmen's compensation laws had been enacted in most of the industrialized countries. Industry became liable for injury to its workers, and obligated to pay benefits during time lost from work, provide medical care, and compensate dependents in case of death. When it was recognized that disablement represented a part of the cost of industrial production and had to be incorporated in the price of a product, management had a strong incentive for preventing accidents. One of the results was the development of safety engineering, which attempts to reduce the number of injuries and disabilities in the occupational setting through removing their causes.

In general terms, safety engineering may be said to attempt to prevent accidents through the elimination of hazards and causation factors, the design of machines and procedures to reduce hazards, and minimization of the effects of hazards by prescription of specified equipment to prevent injury or reduce its severity. Common techniques for achieving these aims may be categorized into three general approaches: 1) advance analysis and critical incidence technique, 2) systems safety engineering, and 3) analysis of design and human limitations. (Hammer, 1972; Smith, 1976; Encyclopaedia Britannica, 1981; McCormick, 1981)

Advance analysis incorporates an operational job analysis which includes a survey of the nature of the tasks, the work surroundings, the location of controls and instruments, and the way an operator performs his or her duties. From this, a functional concept of potential accidents (errors that might occur while the operator is working at the machine) is developed. Such analysis of tasks and working conditions can often be quite revealing. The analysis of "near accidents" is used to reveal how operators react to emergency situations. The knowledge gained in such studies can be used to train operators to anticipate hazardous situations and to avoid them.

The critical incidence technique involves selecting a random sample of workers from various parts of a plant who are exposed to a variety of hazards. An interviewer questions these workers, asking them to describe near-miss safety errors they may have been involved in, or unsafe conditions they may have observed. The data are then classified into hazard categories to reveal factors which have contributed to actual or potential loss-producing events.

Systems safety engineering has received considerable attention in recent years, especially in the aircraft and nuclear power industries. Its primary aim is the reduction or elimination of all hazardous consequences of equipment operations, and commonly involves a systematic application of special anayltical techniques, derived criteria, evaluation methodology, and management skills. Because of their importance in the overall scheme of managing technological hazards, certain of these concepts (i.e., reliability block diagrams, signal flow graphs, minimal cut sets, fault tree analysis, and event tree analysis) will be described in greater detail below.

The analysis of design and human limitations emphasizes the interaction between the human operator and the machine, including the layout of working areas, the matching of equipment to human capabilities, and the design of controls and displays. In many instances an engineer is given instructions to design equipment in such a way as to "prevent reaching beyond the range of vision", to "avoid off-balance positions", or to "provide adequate headroom and clearance" without precise descriptions of these requirements. On the other hand, quantitative information can lead to the "average value" fallacy. Arrangements based on average values would be unsuitable for 50% of the operators in a normally distributed group. Considerable progress has been made in this area, however, especially in terms of the design of controls and displays.

Systems safety engineering is one of the most important developments in recent years with respect to the management of technological hazards. Its primary advantage is that it encourages the engineer to broaden his or her perspective and include many additional factors in the calculations. The general character of systems safety engineering can be illustrated by referring to a few of the more important analytical techniques. (The following paragraphs draw upon: McCormick, 1981; Shooman, 1968; Smith, 1976; and U.S. Nuclear Regulatory Commission, 1975.)

Reliability block diagrams are simple schema used to protray the manner in which a system functions. Such a diagram may be characterized as showing how a signal called "system operation" is transmitted from an input to the output of a system. The system is broken down into units or components which operate either in series or in parallel. The reliability of the overall system can be calculated by aggregating the various reliabilities of the components (which are expressed as probabilities). For example, the respective reliability figures for two units operating in series are multiplied together to give their combined reliability. The reliability of two units operating in parallel is obtained by subtracting the product of their respective reliabilities from the sum of their reliabilities. The combined reliability of systems whose components are sequential or load sharing (only one unit of the system is in operation at a single time) or which are cross-linked so that some components are included in more than one sub-system can also be calculated. Any such system, no matter its complexity, can be decomposed into block diagrams and its reliability thus calculated.

Signal flow graphs constitute an alternative technique for characterizing simple systems which has been extensively developed, in part because of its similarity to circuit theories used in other branches of engineering. The graph for a signal flow analysis consists of a set of nodes and a collection of branches joining the nodes together. Two adjacent nodes represent the input and output of a single component (i.e., each component is represented by a branch in the graph and has a specified direction corresponding to the direction of the signal). The "weight" of the branch represents the event of transmitting the signal through that component. A path is a sequence of branches in which the output node of one branch is the input node of the next. The overall reliability of a system can be calculated by aggregating the separate reliabilities of its component paths in a similar manner as with reliability block diagrams.

A cut set of a system is defined as a set of system events that, if they all occur, will cause a system failure. Although the concept is general, it is most easily visualized as the severing of all branches between input and output nodes in a signal flow diagram. A minimal cut set of a system is a cut set containing system events that are not a subset of the events of any other cut set (i.e., removal of any single system event from a minimal cut set would cause the set to no longer be a cut set, so that the system would no longer fail). Cut sets and minimal cut sets are useful concepts for the analysis of common cause failures (failure of multiple components due to a single secondary event) and common mode failures (common cause failures that act on a set of functionally identical components, thereby causing the failure of duplicate or redundant subsystems). Even though potential common cause or common mode failures are of major concern to the reliability analyst, the rate of systems failure due to them can only be estimated. Of course, any available data should be used, but normally the determination of such failure rates is an exercise in engineering judgment.

Fault tree analysis is a technique by which the relationships between many events that interact to produce other events can be described using simple logical relationships to create a structure representing the system as a whole. The first step is to understand how the system functions. A system function diagram (or flow diagram) is used to describe the pathways by which signals or materials are transmitted between the components comprising the system. Reliability block diagrams are examples of signal function diagrams. But a second diagram, a functional logic diagram, is sometimes required to portray the logical relatiohships among the components. This logic diagram often varies from the functional diagram. A third tool frequently used to understand the overall functioning of the system is a falure modes and effects analysis, which is an inductive anlaysis used for systems for which the expense and effort of a full fault tree analysis are not warranted. Only after the functioning of the system is fully understood should a fault tree be constructed.

In the construction of a fault tree, the system failure event that is to be studied is called the "top event". Successive subordinate events that may contribute to the occurrence of the top event are then identified and linked to the top event by logical connective functions. The subordinate events themselves are then broken down into their logical contributions and, in this manner, a failure event tree is recorded graphically by arranging the events into a tree structure using symbols called "gates".

When a contributing failure event can be divided no further, or when it is decided to limit further analysis of a subsystem, the corresponding branch is called a "basic event". If the basic event for a branch is a basic mode failure such as a structural fault, it is termed a "primary fault". If it represents a subsystem failure due to excessive operational or environmental stress, it is called a "secondary fault". In practice, all basic events are taken to be statistically independent unless they are common cause failures.

Once the tree structure has been established, subsequent analysis is deductive and takes two forms. A "qualitative analysis" is performed to reduce the tree to a logically equivalent form representing the specific combination of basic events sufficient to cause the undesired top event to occur. Each combination will be a minimal cut set of failure modes for the tree. A "quantitative analysis" then transforms the established logical structure into an equivalent probability form and numerically calculates the probability of the occurrence of the top event from the probabilities of occurrence of the basic events. The probability of a basic event is the failure probability of the component or subsystem during the operating time under examination. The failure probability is either constant with time, or changes over time.

For trees containing several tens of events, even when rare-events approximations may be used, computer help will be needed. Certainly for all large trees (e.g., 100 or more basic events), or for the exact calculation of trees containing more than a dozen or so basic events, a relatively sophisticated general computer program is required.

Event tree analysis and decision tree analysis are inductive logical methods for identifying the possible outcomes of a given initiating event. They differ depending on whether the outcomes result from human choice (decision tree) or from the laws of natural science (event tree). An initiating event in a decision tree is a decision, say, to accept a particular risk, and the various outcomes depend upon subsequent decisions. An initiating event in an event tree is typically a system failure, and the subsequent events are determined by the system characteristics.

Once an initiating event is defined, all the safety systems that can be brought to bear subsequent to the occurrence of the event must be identified. After the systems for a given initiating event have been identified, the set of possible failure and success states for each system must be defined and enumerated. Careful effort is required to ensure that potential failure states are not included in the success definitions. Much of this analysis is done with the fault tree technique described above. If bifurcation (two-state) modelling is employed, then one failed state and one sucess state are defined for each system, and each gives a branch of the tree. If a greater number of discrete states are defined for each system (as when including partial failures), then a branch must be included for each state.

Following the proper definition of the system failure and success states, the states are then combined through the decision-tree branching logic to obtain the various accident sequences that are associated with a given event. The system states on a given branch of the tree are conditional on the previous states having already occurred. Thus, it is possible to "prune" an event tree by eliminating all the branches that have a zero probability for at least one event.

A major concern in event tree construction involves the timing of events. In some cases, the logic changes depending of the time at which events take place. Then "phased mission" analysis is needed to model the system that changes during the sequence of events, even though the safety system components remain the same.

Once the final event tree has been constructed so that the results corresponding to each sequence of events have been defined, the final task is to calculate the probabilities of system failure. Fault tree analysis is used to compute the conditional probabilities characterizing the likelihood of each branch of the event tree.

This section describes several approaches in the field of safety and reliability engineering that can contribute to the analysis of technological hazards and the design of appropriate responses to them. Together, these concepts characterize an approach to the analysis of technical and engineering aspects of technological hazards. In particular, safety and reliability engineering can describe the functioning of hazards associated with many kinds of technologies as well as providing estimates of the costs both of the consequences and of preventative measures. They constitute one of the fundamental analytical bases for the management of technological hazards.

Cost-Benefit Approaches

The basic approach of cost-benefit analysis is one of the most intuitive ideas of human behavior: before undertaking an activity, one should compare the costs of doing so with the benefits one expects to derive from it--as long as the benefits outweigh the costs of obtaining them, the activity is considered rational. Of course, this basic idea underlies much more than just cost-benefit analysis. It is also the basis for many accounting procedures in business and the simple budgeting of an individual household. The approach can also be used to determine the level of activity to undertake under conditions of limited resources as well as whether or not to pursue that activity at all. (Mishan, 1976; Irwin, 1978; Hofmann, 1981)

Cost-benefit analysis is commonly used in investment decisions. Engineering analyses usually include costs of construction and operation, and accounting provides detailed information on costs and return on investment. But the perspective of such engineering and accounting analysis is normally that of the individual or enterprise contemplating the investment. Cost-benefit analysis extends the examination to a much broader perspective. It is a common notion that what counts as a benefit or a loss to one part of the economy--to one or more persons or groups--does not neccessarily constitute a befefit or loss to the economy as a whole. Cost-benefit analysis aims at evaluating the consequences for the economy as a whole.

The fact that the activities of a private enterprise are guided by the profit motive does not mean that the enterprise does not produce benefits for other parts of the economy. The "invisible hand" of economic motivation directs such self-interested activities in the business world to produce benefit, not only for the employees of the enterprise and those with whom it does business, but also for the society as a whole. It is possible, in fact, to define simple and sufficient conditions under which the uncompromising pursuit of profit acts always to serve the public interest. These conditions can be reduced to two: that all effects relevant to the welfare of all individuals be properly priced on the market (i.e., reflect all costs), and that perfect competition prevail in all economic activity. (Mishan, 1976, p. xi) Once these ideal conditions are compromised, however, the provision of goods and services and the determination of the prices thereof in the marketplace may not serve the public as well as some other set of outputs and prices.

The appraisal of a project using a cost-benefit approach does not involve questions much different from those asked by engineers or accountants. The essential difference is that the same sort of question is asked about a wider group of people and is asked more searchingly. Broadly speaking, the economist substitutes the concept of "social benefit" for the more precise concept of "revenue to the firm". For the concept of "cost to the firm", "opportunity cost" (the social value foregone when resources are devoted to the project in question rather than to some alternative) is substituted. Finally, the concept of "excess social benefit over cost" replaces the "profit to the firm". (Mishan, 1976, p. xii)

Use of the cost-benefit approach usually implies application of a concept of social betterment that amounts to a potential Pareto improvement. The central role of this concept in the approach warrants its description in some detail.

(Mishan, 1976, pp. 391-392) Suppose CV is the sum of money which, if received or paid after the economic exchange in question, would make the individual no better or worse off than before the change. If, for example, the price of loaf of bread falls by ten cents, CV is the maximum sum an individual would be able to give up and still satisfy his or her need for bread to the same degree as before. Conversly, if the price of bread rose by ten cents, CV is the minimum sum an individual would have to receive to continue to feel as well off as before. In general, some people gain and some people lose in any economic change. The CVs of all those who gain (the largest sums they would be willing to give up--a positive amount) may be added to the CVs of all those who lose (the smallest sum required to make them feel as well off as before--a negative amount). If the resulting algebraic sum is positive, the benefit to those who gain is greater than the loss to those who lose, and a Pareto improvement can be realized. If, however, the algebraic sum is zero or negative, the change in question does not represent a potential Pareto improvement. Furthermore, the magnitude of the positive algebraic sum indicates the extent of the potential Pareto improvement, while the magnitude of a negative algebraic sum measures the extent of the potential reduction in welfare.

In applying cost-benefit analysis, however, one must move from the general concept of CVs to specific estimation of benefits and costs. The transition can be made following the conventions of the market economy, with people regarded as both producers and consumers. If, as producers, people are no worse off in consequence of some contemplated change but, as consumers, are better off, then on the whole they are better off and a Pareto improvement is achieved. Or if as consumers people are no worse off, but as producers are better off, then again a Pareto improvement results. Of course, if people are better off both as producers and consumers a Pareto improvement is also achieved.

It should be noted that such calculations are to be interpreted as abstract contributions to potential situations of Pareto improvement. A cost-benefit analysis which shows a gain of $100,000, for example, does not mean that everyone actually is made better off in an amount equal to $100,000 through costless redistributions. To return to the earlier illustration, Pareto improvement is actually realized only if the individual in fact receives or gives up the sum which counteracts the change in the price of a loaf of bread.

The use of the cost-benefit approach is widespread. (E.g., Mishan, 1976; Luft, 1976; Barum, 1980; McCormick, 1981) But despite its considerable use, there has also been a great deal of criticism of its application. In reviewing the use of the approach in decisions about health, safety and enironmental regulations, for example, Barum (1980, pp. 481-492) identifies the folowing difficulties in its application: 1) inadequate identification of costs and benefits of proposed actions, 2) difficulty in quantifying some effects, 3) difficulty in determining discount rates for valuing future effects in present analyses, 4) improper distribution of costs and benefits, 5) promotion of self-interest and other analytical biases, and 6) special problems of accountability. A brief description of these criticisms is perhaps a good introduction to the issues involved in the use of cost-benefit approaches in practical situations.

The problem of inadequate measurement of attributes is related to deficiencies in forecasting technique. There is often simply no information available upon which to base the necessary analysis. To take an example from the production of chemicals, until adequate toxicological studies have been performed, it is impossible to be certain about the effects of a substance to be marketed. Further, many environmental effects, such as alteration of ecosystems, cannot be estimated with confidence because no accepted method exists to measure these effects. Nevertheless, guidelines for structuring the attempt to measure attributes do exist. For example, one scheme of classification includes the following: internal versus external effects, incomensurables (which cannot be translated into a common denominator) and intangibles (which cannot be measured at all), and direct versus indirect effects. (Sassoone & Schaffer, 1978, pp. 32-40) The most important element of a cost-benefit analysis is the identification of relevant attributes; but even when costs and benefits are identified, they might be excluded from the analysis for pragmatic reasons. They may be too costly or too complex to measure, or they may raise issues that would stimulate considerable opposition to the organization performing the analysis should their inclusion become known.

Cost-benefit analysis works best when there is a widely accepted method to measure the costs and benefits, and when the resulting measurement can be expressed in monetary units. But all too often, the consequences of the project being examined include effects for which there are no accepted prices. This is nowhere clearer than when the analysis must evaluate the effort to save human lives (i.e., avoid early death). There are three common approaches to assigning value to human life. (Luft, 1976, p. 448) The first two are related to the productivity of the individual in question. The first uses an estimation of the expected earnings of the individual to approximate his or her marginal productivity. The second approach starts with this crude measure of productivity, and subtracts the estimated marginal consumption of the individual to approximate the net surplus available to society. The third approach is based on the value of a "life" rather than of "livelihood". The direct valuation of a "life" is produced by asking people how much they would be willing to pay to reduce the probability of death. Not surprisingly, quite different values emerge in different cost-benefit studies. The U.S. Environmental Protection Agency Office of Radiation Programs establishes its standards at levels that will not cost more than $500,000 per life to be saved; the U.S. Consumer Product Safety Commission uses values ranging from $200,000 to $2,000,000 per life in its analyses. (Barum, 1980, p. 485) Such variation suggests that the measure of the value of life incorporates other factors than merely the economic value of the individual. Although public officials generally deny that they are making arbitrary judgments in their analyses, the variation in such estimates in different studies suggests that the value used in any single study an implicit, and perhaps unavoidable, choice.

"Shadow prices" are used in cost-benefit analysis whenever the attribute under consideration is not bought and sold, so that it has no market price, or when the market price does not adequately reflect the true social value of the attribute. A shadow price may be defined as a value associated with a unit of some good which indicates how much a specified index of performance can be increased (or decreased) by the use (or loss) of the marginal unit of that commodity. (Sassone & Schaffer, 1978, p. 50) However, even where market prices are felt to misrepresent the true societal value of the attribute in question, an analyst is often well advised to use the market price and note the direction of the bias as he or she perceives it. This is because there is a large element of subjective judgment in the determination of shadow prices, and their use can sometimes call the credibility of the analysis into question.

The discount rate for valuing future costs and benefits in present analyses is another chronic problem. The discount rate is a way of expressing preference for current actuality over future potential. The further into the future an effect is to occur, the greater the reduction of its magnitude in the current analysis. However, no rate of return has been developed which adequately measures the intertemporal preference of society as a whole. A high discount rate penalizes effects far in the future (reduces the magnitude of benefits and costs alike in the contemopary decison) and a low discount rate increases the relative value of impacts further into the future. The choice of a discount rate should be a deliberate attempt to apportion the costs and benefits among groups and generations, according to the values predominating in society. (Stokey & Zeckhouser, 1978, p. 173) However, there is no agreement as to how to accomplish this. One argument is that the dispersion of social benefit over a large number of individuals means that the "social discount rate" should have a lower value than the "individual discount rate" (i.e., that future costs and benefits are of greater value to society as a whole than to individuals). But a counter argument holds that social projects should be charged their true opportunity costs, which would call for a higher discount rate and greater discounting of future effects. One compromise would be to use the lower social discount rate, and to reflect the opportunity costs in the valuation of the attributes. But neither parliaments nor courts, the two traditional sources of value determinations, have succeded in either defining a generally applicable discount rate or in specifying how a discount rate should be determined in a specific analysis.

The differential distribution of costs and benefits, and the resulting need for transfer payments or "costless" redistributions, has already been mentioned in the introduction to this section. Applications of the cost-benefit approach sometimes include attempts to account for unequal distributions of the costs and benefits. However, it is not only diffucult to determine the distribution of the effects over various groups, but even more so for the distribution over succeding generations. No matter how careful and searching the analysis, there will always be uncertainties and questions. But in effect, the distribution of costs and benefits over groups and over time each introduce ethical issues which transcend both economic and legal perspectives. The assumptions incorporated in analyses are necessarily inadequate, precisely because they are made by the analyst rather than by society. At best they are approximations which depend on the skill and perception of the analyst producing them.

This suggests that analysts can easily play a "numbers game" with cost-benefit analysis. The purportedly objective framework of cost-benefit analysis can be used to promote a particular position rather than to analyse the options. Through manipulation of the discount rate, selection of arbitrary values for specific effects, exclusion of selected effects, or self-serving assumptions about distributional fairness, the analyst can bias the results in ways that are not easy to observe. Like most tools of social and political analysis, cost-benefit anaysis can be misused if the analyst so desires or is careless.

The possible misuse of cost-benefit analysis introduces the problem of accountability. The danger is that the apparant objectivity of the analysis might be used to justify choices that are in fact questionable. Thus, any such analysis should be exposed to review by analysts other than those who have performed the work, and a clear specification of the assumptions and judgments which have been incorporated should be a part of the analysis.

The intent of reviewing these difficulties in the application of cost-benefit analysis has not been to challenge the credibility of the approach. Rather it has been to insure a realistic recognition of the possibilities and limitations. The underlying idea--the comparison of the advantages and disadvantages before acting--is so basic that it must be used in some form. But it is important to recognize that proper use of this simple idea is complex and difficult. The cost-benefit approach is one of the fundamental tools in the examination of alternative responses to technological hazards and their likely consequences. It is and will remain a key approach in the evaluation of hazards and response programs, and as such is one of the principal elements of the management of technological hazards.

Analysis from the Perspective of Society: Technology Assessment and Environmental Impact Analysis

Just as the idea of comparing the expected costs and benefits of a contemplated action is the fundamental basis for many analytical studies, the idea of examining the set of alternative ways of accomplishing the desired end serves as the guiding principle for many evaluations of responses to technological hazards. Cost-benefit analysis constitutes the professional, analytic manifestation of the basic economic comparison, and technology assessment that of the comparison of action options.

In most general terms, technology assessment consists of four components: analysis, forecasting, evaluation and recommendation. (Battelle, 1976) The approach may be seen to encompass a range of evaluations from broad problem-oriented studies, which may examine a major societal problem (e.g., energy shortage), through technology-oriented assessments intended to anticipate the effects of a developing technology (e.g., the videophone), to the most specific form, the project analysis, which focusses on a single application of technology (e.g., the siting of a power plant). (Porter et.al. 1980, p.3) Two common types of studies can represent this whole range of evaluations: technology assessment and environmental impact analysis. Both strive to systematically examine the effects of technology on society, and both give special emphasis to those consequences which are unintended, indirect, or delayed. However, there are also differences between the two. Technology assesment usually deals with a technology that could be geographically situated almost anywhere, whereas environmenal impact analysis is more likely to consider the technology in terms of a specific site. Technology assessment is more likely to explore a wide range of possible subsidies, incentives, regulations and so forth. Environmental impact analysis more often focusses on "go or no-go", "build or not build" decision alternatives for a particular project. (Porter et.al. 1980, pp. 3-4) The differences, however, are not always distinct in practical applications. The two will be treated here as closely related manifestations of a common approach, and will be designated by the abbreviation "TA/EIA".

It can be argued that no methodological framework can be comprehensive enough to apply to all situations. Nevertheless, the approach of TA/EIA calls for the integration of a variety of kinds of information and expertise from different disciplines. This requires a particularly clear methodological treatment. Figure 3.1 describes a systematic approach to the social assessment of technology which has exerted considerable influence over the general development of TA/EIA. This system can be described as a threefold approach which, through an iterative procedure incorporating feedback of data and results, integrates the processes of analysis, decision making, and communication of information into a single system. (OECD, 1975, p. 19)

Porter and his colleagues, in an extensive review of the field, characterize TA/EIA as follows (Porter et.al. 1980, pp. 4-5):

> A TA/EIA should begin with a careful definition of the problem. It must accurately describe the state of the art of the relevant technology and forecast the changes that the technology is likely to undergo. It should also describe the relevant features of the society, its values, and the ways of managing the technology, and should anticipate how these factors might limit and shape the context in which the

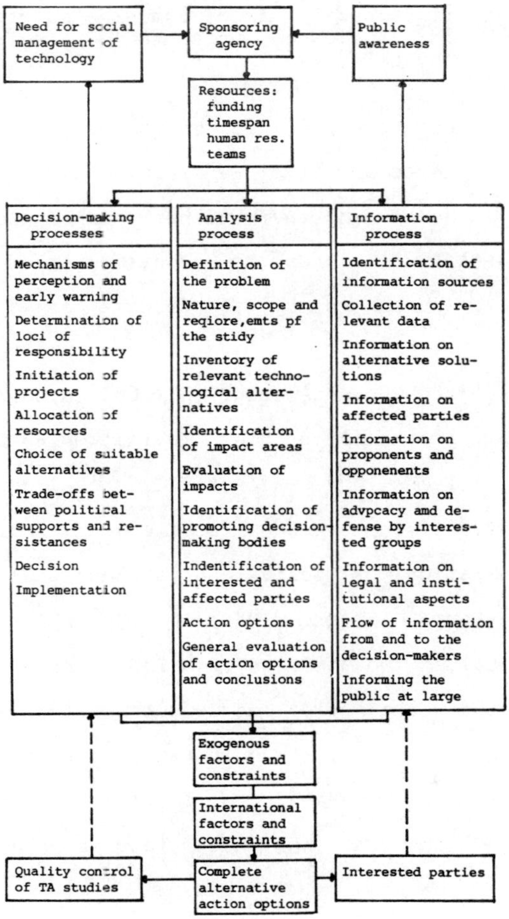

Figure 3.1 A System for Social Assessment of Technology
Source: OECD, 1975, p. 20

> technology must function in the future. It must identify areas of the social and physical environments that are likely to be impacted by the technology, and analyze the character and extent of significant impacts. Policy options open to decision makers for the modification of expected impacts must also be identified. Finally, since no TA/EIA can be effective unless its results are communicated to the user, it should be presented in a clear, logical and effective manner.

This description both gives a quick characterization of the scope of TA/EIA and suggests how it differs from risk analysis. TA/EIA encompasses all human and environmental consequences with equal weight. Risk analysis tends to emphasize those consequences which lead to human injury or disease or damage to the environment.

Following the above description, the components which make up a TA/EIA are: 1) problem definition, 2) technology description, 3) technology forecast, 4) social description, 5) social forecast, 6) impact identification, 7) impact analysis, 8) impact evaluation, 9) policy analysis and 10) communication of results. (Porter et.al. 1980, pp. 54-60) These various elements are called "components" of a TA/EIA rather than "steps" to emphasize that the activities do not follow a simple, linear sequence. Although their listing in this order follows a logical progression, it is important to iterate or repeat various components based on what has been learned in doing the other tasks.

The first component in a TA/EIA involves the determination of the nature and scope of the study. The breadth and depth of the expected analysis should be estimated, with reference to the objectives, or questions to be answered, as well as the available resources. A preliminary assessment can help determine the limits of the study. One important aspect of problem definition should be the identification of the parties at interest to the assessment subject and the nature of their interest. This will be important in helping to identify the social and political values that might be included to help define the important impact and policy sectors to be examined.

A thorough and accurate description of the technology under assessment is crutial. Particularly in a technology-oriented assessment (as opposed to a problem-oriented assessment), this second component can make up a considerable portion of the overall activity. A description of a technology should include the identification of the major technical parameters, alternative ways in which they can be implemented, competing technologies which might serve as alternatives, and a characterization of the applicable technological delivery system (the pattern of information exchange, flow of materials, and exercise of authority among the various institutions and organizations involved in the development and use of a technology).

The third component, technology forecasting, attempts to anticipate the character, intensity, and timing of changes in technologies. Within TA/EIA it is used to characterize the future of the technology under assessment. It can identify key uncertainties, potential breakthroughs, upcoming substitutions of one technology for another, likely cost reductions, and new applications.

The next component, social description, should characterize the state of society in terms of interactions with the technology in question. Both the determination of parties at interest and the characterization of the technological delivery system can help identify such interactions.

Social forecasting seeks to represent the most plausible future configurations of relevant dimensions of society or to project changes in such social parameters. Provision of a range of qualitative, alternative futures in the form of scenarios (expositions of possible future developments or conditions) is a typical approach.

In TA/EIA, "impacts" refer to the products of the interactions between a technology and its social context. An important attribute of TA/EIA is the identification of both direct and higher order impacts (direct impacts are those directly attributable to the technology; higher order impacts are those which derive from the direct impacts). Classification of impacts is usually either along disciplinary lines (e.g., environmental, psychological, institutional/political, social, technological, legal, economic) or according to the various parties affected.

Impact analysis examines the likelihood and magnitude of the impacts identified. The impacts identified in the previous component must be characterized in ways that capture both the seriousness and the likelihood of their consequences.

Impact evaluation attempts to determine the interrelationships and significance of the impacts in terms of societal goals and objectives. Impact evaluation involves judgment about values and goals which may not be clearly defined. It is an activity which must be the responsibility of those performing in the TA/EIA alone. If the responsibility is to be spread, then a greater number of people must be involved in the evaluation process. For this reason, TA/EIA applications often encourage the active participation of lay people and representatives of interest groups. Impact evaluation procedures should be as clear as possible to allow the user to accept or reject the conclusions on the basis of information rather than inference.

Policy analysis relates the impact assessment to the concerns of society, especially so as to place the conclusions in a proper perspective with respect to relevant programs of action. The policy analysis compares options for the implementation of technological developments and for dealing with their desirable and undesirable consequences. Explicit recommendations may or may not be appropriate, depending on the preferences of the sponsors, users, and performers of the TA/EIA.

Careful attention should be paid to communication of the results. Determination of the appropriate study presentation depends on the aims of the study, the sponsor's desires, and the resources available. Private sector assessments may or may not include this aim.

Technology assessment and environmental impact analysis represent a very broad approach to the examination of alternatives options for action. The approach has most often been used in single, one-time studies. TA/EIA as an approach has seldom been applied in the context of a continuing program of evaluation and management of technological hazards.

The TA/EIA approach as described in this section is in general broader than the approach to the management of technological hazards employed in this study. This is because the former places greater emphasis on the anticipation of future events and conditions. This is not to say that the approach in this study does not include consideration of consequences in the future. Of course it does. Rather, it is merely to highlight a difference of emphasis. The TA/EIA approach gives special emphasis to consequences which are unintended, indirect and delayed. The approach in this study gives greatest emphasis to consequences related to human injury and health and to environemntal damage. But despite these differences in emphasis, the general approach of TA/EIA is applicable to the subject addressed here; problem definition, analysis, evaluation, and recommendation are all key elements of the management of technological hazards as it is described in this study.

One of the more imporant aspects of the TA/EIA approach which is also of utmost importance in the management of technological hazards is the identification of relevant social actors. The TA/EIA methodology does this principally through the examination of what is called the technology delivery system, the analysis of the parties at interest, and the performance of policy analysis. Although these concepts will not be used directly in the remainder of this study, they have a great deal of similarity with those that will be.

From the Perspecitve of the Individual Firm: Risk and Insurance Management

The idea that the control of hazards should be an integral part of the management of a firm has been recognized by students of management at least since 1916, when Henri Fayol included "security' (protection of property and persons) together with production, sales, finance, organizing and direction, and accounting as the essential "organic functions" of management. (Head, 1982, p. 181) Of course, recognition of this function takes different forms in different firms. But because one of the traditionally more important strategies for dealing with damaged property or injured persons has been insurance against loss, the anticedents of this activity today are to be found in financial management, and in insurance management in particular.

"Risk and insurance management is merely the application of general concepts in scientific management to a particular problem." (Olson & Simkiss, 1982, p. 114) This management approach is commonly understood to be the "planning, organizing, directing, and controlling of the resources and activities of an organization to minimize, through effective loss control and financing, the adverse effects of potential and actual losses upon the organization." (Head, 1982, p.181) Traditionally, risk and insurance management has focussed on measures aimed at avoiding or reducing the probability and/or severity of the occurrence of losses (physical loss control), and on provisions for financing losses that do occur (financial loss control). (Carter, 1982, p.5) Thus, the emphasis in risk and insurance management has been on what is called "pure risk", where no possiblity of gain exists (e.g., fire, theft, negligence, injury or death). The realm of "speculative risk", which offers both the chance of gain and loss (e.g., changes in technology, in consumer taste, in government regulations, or in market prices), has not usually been within the scope of the risk manager's activities.

Risk and insurance management has come to be viewed as more important in the overall management of the firm as it has been recognized that pure risks can involve substantial losses, sometimes even large enough to threaten the survival of the firm under conditions where profits and liquidity are limited. In simplest terms, the task of the risk manager in the firm is to increase the wealth of the owners or shareholders by selecting strategies that entail the optimal combination of expected return and risk. "The decision about whether to minimize financial disruption through the purchase of insurance, or, conversly, to internally assume the exposure to loss should be made on the same basis as any financial decision such as the investment in plant or equipment, whether to buy or lease, the choice of securities in an investment portfolio, and so on." (Olson & Simkiss, 1982, p. 115)

It is commonly argued that expenditures for physical loss control should be pursued so long as benefits exceed costs. (Green & Serbin, 1978) However, there are considerable difficulties in realizing this aim. The most important considerations are the flow of costs and benefits over time, and the uncertainty regarding outcomes. These can be handled by the same techniques that are used for the evaluation of other types of investment decisions (see, for example, the section on cost-benefit approaches in this chapter). But equally important in certain circumstances is the estimated impact on the size and variability of individual and aggregate losses in any one time period. Thus, more resources might be expended on eliminating the possibility of suffering very severe losses, even though larger savings in terms of annual loss expectancies might be realized through attending to potential loss-producing events with higher expected frequency but lower individual loss potential. (Carter, 1982, p. 6) (For a theoretical justification of such choices see the section on cognitive attributes of decisions about risk in Chapter 5.)

Similar considerations apply to the alternatives for financing losses. Losses can be financed through three basic strategies: charging losses to current operating costs, ex ante anticipation of costs through purchase of insurance or setting aside internal funds, or ex post financing through borrowing. (Carter, 1982, p. 6) However, the evaluation of outcomes under different strategies is not simple. The fixed premium insurance strategy is straightforward: barring problems involving the financial integrity of the insurance company, there is no range of variation. The outcome is the known cost of the premium. However, determining the range of outcomes under a strategy of self-insurance is quite complicated. In order to evaluate the possibility of financial disruption it is necessary to assess the likelihood that each of the possibilities in the range will actually occur. One the one hand, the range of outcomes might be narrow and the expected result very likely to occur. But on the other hand, the range of possible outcomes could be extremely wide and the expected result little more likely to occur than other outcomes. (Olson & Simkiss, 1982, p. 116) Thus, trade-offs between and among risk handling costs, financial security, and cash-flow effects are important.

> In pursuing such objectives, a risk manager is constrained by the need to keep retained losses within certain limits, which might be laid down in his own or other departmental budgets. Moreover he must pay regard to the impact of retained losses on the company balance sheet and consequent stock market reactions as reflected in movements of its share price. If a firm could work to an infinitely long planning horizon, then it would never pay to insure because to do so increases annual expected risk costs and thus reduces expected profits, but in the real world liquidity, credit-worthiness and share price considerations focus attention on actual annual risk costs. (Carter, 1982, p. 6)

If a firm is to avoid increases in the possibility of liquidity shortage, decisions to retain losses without any funding provisions must take into account monthly fluctuations in both direct losses (e.g., expenditures for repairs, medical treatment, liability) and indirect losses (e.g., reduced income during down time, adverse effects on market position or reputation), as well as the normal fluctuations resulting from operating cycles, seasonal variations, and so on. Large reductions in liquidity, abnormal decreases in cash flow, and the like can be handled by carrying cash reserves or by borrowing. The former is costly in terms of earnings foregone. The management of cash and working capital in general involves balancing risk (both pure and speculative) and profitability. The size and nature of loss producing events can also affect a firm's ability to finance losses by borrowing. Additional borrowing after a loss has occurred adds to the financial strain at a time when earnings may already be impaired.

It is for reasons such as these that some risk managers pay special attention to those loss producing events which may disrupt earnings and thereby cash flow, especially in considering physical loss controls. A major liability claim, providing it does not adversly affect sales, is more easily financed than a major interruption of business production of equivalent monetary value. One of the paradoxes of risk management within the firm, or at the level of society for that matter, is that the magnitude of the potential loss is generally inversely related to the ability to predict the occurrence of the loss producing event. Thus, the conduct of risk and insurance managment is of necessity more an art than a science.

Nevertheless, there is widespread agreement as to the fundamental procedures which are required in risk and insurance managment. Although there are variations in details (see, for example: Head, 1982), the specification by Oslon & Simkiss (1982) is typical. They list the following procedures, in chronological order:

1) identification of the pure risk events that may impose financial losses on the enterprise

2) evaluation of the potential costs that would be involved if such events should occur

3) estimation of the frequency with which such events might occur

4) analysis of the alternatives that are available

5) selection of the most desirable alternative in light of the overall goals of the enterprise at the time the decision is made

6) periodic re-evaluation and monitoring

The similarity between these procedures and those of other approaches described in this chapter is obvious. However, the particulars of the perspective of the individual firm are clear not only from the manner in which these basic procedures are formulated, but also from the manner in which they are pursued.

Olson & Simkiss (1982) characterize the identification

process as the most important for the risk and insurance manager, simply because the failure to recognize one or more potential loss producing events can result in financial disaster. They suggest several approaches, including a systematic evaluation of the financial and operational characteristics of the enterprise, incorporating its relationships to suppliers and users as well as other elements in its operating environment. Such checklist approaches require the investigation of alternative sources of supply, of distribution, of sources and uses of funds, of financial capacity, and of operations used by the firm. But despite the best efforts of the risk manager, important aspects can be overlooked. Social, economic or technological changes can result in new risks. Sometimes such developments are such that there is little experience within the firm, or even among other enterprises of a similar nature. Furthermore, the legal standards by which enterprises are judged, including the circumstances under which they can be held liable and the costs that may be imposed on them, are in constant change. Therefore, the identification of risks to the firm must be a continuing process.

Following an identification of the pure risks facing the firm, the next step is to estimate the costs that might be incurred should those events take place. By performing this function second, unimportant events can be eliminated from the process as early as possible. The best way to do this is to determine whether the impact of the events is of importance. There are two major difficulties in this determination. First, two or more events may occur simultaneously, and although the financial consequences of the individual events might alone be unimportant, taken together the impact prJ be serious indeed.

For exapvui; a warehouse fire includes not only the loss of the structure, but also the goods stored there. If this loss of inventory so depletes stocks that deliveries pmtg be postponed, this adds an additional cost. Any one of these alone pn′"g be easily prdr′il, But responding to all three at the sapi time might be much more demanding. The second difficulty is that events can give rise to indirect as well as direct costs. Legal liabilities are good examples. Claims against alleged wrongdoing involve the cost not only of the reparation paid to the victim, but also the costs of the settlement process, of eventual product recall, and of negative pre)ig impact due to notoriety deriving frop the successful claim. Indirect costs also derive from property losses. The most important is probably interruption of business activities resulting in serious reduction of revenue.

After identifying the potential pure risk events and their probable financial impact, the third step involves estimating the frequency with which such events might occur. The major difficulty in this is lack of information. There is no readily available source of information which may be used to make such estimations for most enterprises. This is partly due to the fact that pure risk events keep changing. But it is also true that the relevant data is seldom collected. The importance of the estimation of frequency can only be understood when it is recognized that an appropriate evaluation of alternative responses is dependent on that estimation. Therefore, those enterprises sophisticated in risk and insurance prdr′ipidg establish procedures for recording their own experience. If the firp is large enough to produce a set of data that is statistically reliable and meaningful, this can be extremely valuable. From the estimates of frequency and the estimates of the potential costs that would result if the events were to occur, it is possible to develop several important pieces of information. For example, it is possible to estimate the likely level of retained losses that would occur using various forms of self insurance, and to compare this with the premiums charged by commercial insurance companies.

In general terms, there are four alternative responses that are usually considered in risk and insurance management: i) deterence, ii) transfer (usually by contract) to some other entity through some mechanism other than insurance, iii) assumption of the risk (usually referred to as self insurance), and iv) purchase of insurance from a commercial insurance company.

"Deterence" refers to programs attempting to reduce the firm's exposure to pure risks. They are often called "loss prevention programs" and may attempt to reduce the frequency with which loss producing events occur, to reduce the financial impact on the firm, or both. Loss prevention programs have been developed for each of the different kinds of pure risks commonly experienced by firms, including occupational disability of employees, damage or conversion of property, legal liabilities, and nonperformance by independent contractors and employees. Occupational safety programs include both reducing the likelihood of accidents through improved design of machinery and the workplace, and educational programs for employees regarding their exposure to dangers. General maintenance of equipment and facilities, security programs, fire detection and fire fighting equipment, and training employees in fire prevention are typical of efforts to protect property. Selection of construction materials or even location can also contribute to the reduction of pure risks.

Loss prevention programs for the exposure to legal liabilities are generally more complicated. For firms for which the general public is frequently on the premises, consideration of the maintenance of grounds and facilities in a safe and secure manner is foremost. For firms producing consumer goods, programs designed to detect defects, communicate adequate warning regarding the use of the product, test the products for unforeseen hazards, and facilitate the recall of the product should it be deemed necessary are typical.

In many countries, the decision of whether to implement certain loss prevention programs is often removed from the perogative of management. This is because the benefits that can be realized by such preventative efforts extend beyond the firm itself, often with benefits to society as a whole that greatly outweigh the reduction in the potential losses carried by the firm. Thus, rational decision making by the firm would produce less "protection" than that which would be considered optimal from the overall perspective. Imposition of loss control programs is most common in the areas of occupational disability and legal liability. Insurance companies also sometimes provide loss prevention assistance as an additional service, and may grant concessions in the premium charged when optional loss prevention programs are implemented.

There are a number of opportunities for a firm to transfer the burden of pure risk events to some other entity using a mechanism other than insurance. Such transfers are possible in principle whenever the firm enters into contractual aggreements with outsiders for either the sale or purchase of goods or services. Such transfers may involve responsibility for loss regardless of cause, or may be limited in some way. Such contractual arrangements are in general extremely complicated, and easily subject to dispute. Their identification and evaluation in the risk and insurance management process is quite difficult.

Two questions need to be addressed concerning such transfers. The first is: why would a firm be willing to assume additional obligations for pure risk events? The tradtional answer is that the party is at a competative disadvantage, and is forced to assume such responsibility in order to engage in the activity at all. However, it is possible that one of the parties is distinctly better able to bear the risk in a cost efficient and expeditious manner. For example, the seller may be in a better position than the buyer to prevent damage to goods during shipment because of his knowledge of the hazards or his relationship with shippers. As long as the additional consideration the buyer is willing to provide for this service is greater than the cost incurred by the seller, the arrangement is mutually advantageous. This leads directly to the second notion: how is the desirability of a contractual transaction to be evaluated? One common rule is that the responsibility should be carried by that party which has greatest control over the occurrence of the event. However, there are many exceptions to this rule.

In general terms, self insurance occurs whenever the costs attributable to pure risk events are absorbed directly by the financial resources of the firm. There are basically three methods of financing self insurance programs: i) an unfunded program with the ensuing costs being taken from cash flow, ii) an internal funded program with costs being taken from working captial or earned reserves, and iii) an external funded program using a trust or other mechanism. The principal advantage of an unfunded program is its simplicity. It is most appropriate when the costs arising from the pure risk events are relatively stable and not large, unless the firm is extremely large. When this is not the case, an alternative is to set aside funds which would be available to use whenever an event occurs. While it is not necessary for such funds to be in the form of cash, it is necessary that the investments or funds be liquid and that their value be secure. The fact that the investment must be both liquid and secure means that the firm will incur some opportunity cost in a funded program. Among the alternatives, captive insurance companies have been the most popular external mechanisms for funded self insurance programs. The captive assumes responsibility for the pure risk events of the parent and purchases protection from reinsurance companies. For the most part, the advantages of this approach lie in greater flexibility in the reinsurance market than in ordinary commercial insurance, in the retention of the profit portion of the premium within the parent firm, and in the application of the insurance expertise gained to the loss prevention program of the parent firm.

Commercial insurance companies have responded to the trend towards self insurance in a variety of ways. They have developed programs designed to more accurately reflect the actual exposure of the individual firm, and to provide ancillary services. In fact, it is now possible to purchase such services (e.g., safety engineering, claims investigation, settlement) on a fee-for-service basis without purchasing insurance.

Although administrative costs, premium payments, and other known charges under different programs may vary widely in both timing and amount, the most difficult problem in risk and insurance management is the choice between the risk and the anticipated savings. Often the alternative which offers the possibility for greatest savings also entails the greatest possibility that these savings not only will fail to materialize, but that the enterprise will suffer serious financial disruption. It is extremely important to identify and evaluate the consequences of each strategy.

Oslon & Simkiss (1982, p, 128) summarize the overall position of risk and insurance management as follows:

> In actual practice, the integration of risk management decisions with the overall goals and position of the enterprise is generally provided by rules of thumb indicating that the program should limit the possibility of uninsured losses during a particular time horizon to some stipulated fraction of revenues, working capital, earnings per share, or other financial measure. Evaluating alternatives to determine whether they conform to these restrictions is not a simple exercise.

Perhaps the most important point to note about the relationship between risk and insurance management and the management of technological hazards as defined in this study is that the two approaches deal with the same or similar problems from somewhat different premises. Risk and insurance management takes the perspective of the private enterprise: its principal aim is to balance the costs of preventing loss producing events within the firm against the potential losses that might be incurred. In this, the legal provisions of liability, workmen's compensation, and the like are important but exogenous considerations. The more general approach to managing technological hazards used in this study, like the approach of cost-benefit analysis, takes a broader aim: the balancing of overall gains and losses of specified target groups (e.g., workers, pregnant women, the elderly), or of society as a whole. The legal provisions which are external to the field of choice available to the private enterprise, and especially changes in such provisions, are very real alternatives in such considerations. Thus, even though the procedures used in identifying, estimating, and evaluating the various elements are similar, these different premises generally lead to different weighting of various factors. Furthermore, common sense (and sometimes legal prescription) limit the scope of the response programs considered by the firm to those activities associated with its employees and its facilities or products. The approach to the management of technological hazards advocated in this study, in contrast, explicitly examines ways of motivating different legal entities and groups to undertake appropriate response. Nevertheless, the typical approaches and techniques used within the industrial firm constitutes important input to the study. The majority of modern technological hazards result from the activities of such enterprises, and it is essential that the attempt to be rational and intentional in responding to such hazards take cognizance of their treatment within such firms.

Conceptual Limitations to Formal Analysis

Several difficulties confronting the formal analysis of technological hazards have been referred to, either explicitly or implicitly, in this chapter. For example, the section on safety and reliability engineering refers to the complexity of fault tree calculations and to the necessity of exercising "engineering judgment" at various points in such analysis. The section on cost-benefit analysis describes a number of difficulties confronting that approach together with responses to them. The most important being the difficulty of identifying and quantifying attributes of a program, of selecting discount rates, and of dealing with distributional effects. Both the sections on technology assessment and on risk and insurance management refer to the scarcity of relevant data, the necessity to perform judgments in the interpretation of available information, and the difficulty of estimating the responses of various parties.

All these difficulties can be characterized as "internal" to the analysis. That is, they arise with respect to the performance of the analysis, and are connected in some way to the procedures used. This does not necessarily mean that the consequences of such difficulties are internal to a study. To the contrary, the resolution of such difficulties within a particular analysis is often viewed by interested parties as arbitrary, even capricious, becoming a source of conflict with respct to the acceptance of the conclusions.

This section, however, raises a different sort of issue concerning the conceptualization of such analyses, difficulties which challange the very idea of performing an analysis. In this sense, the problems presented in this section may be thought of as "external" to the procedures of a given analysis. Some of the internal or procedural difficulties mentioned above are related to the external problems, so that some of the ideas presented here will already have been mentioned in other sections. But the discussion here concerns more whether or not such an analysis should be performed rather than how it should be carried out.

The first point is that comprehensive analysis of technological hazards and their social and environmental consequences is theoretically impossible. (This point is developed with respect to general planning procedures in: Ronge & Schmied, 1972.) This has obvious impact on the procedures employed in formal analysis. The engineer reaches practical limits in tracing sequences of events when constructing event trees or fault trees, and often must stop the examination and treat the remaining sequence as a "base event". Whether his judgment is correct and the remaining branches are insignificant cannot be evaluated a priori. Some research suggests that certain bias may be found in such judgments. (Fischhoff, Slovic & Lichtenstein, 1978)

Similarly the analyst performing a technology assessment must decide which social groups are relevant to the evaluation, as well how to approximate their views in the analysis. In all but the most limited cases, it is generally impossible to include reference to all groupings of individuals that might be affected. Yet individuals are likely to respond differently within various groups or settings in which they find themselves. Whether or not the analyst in fact includes the important groupings and excludes the unimportant ones is extremely dificult to determine. It is equally difficult to know if the risk manager in a firm has identified all important "pure risk events" confronting the enterprise. In fact, this point was characterized as the greatest uncertainty in that entire approach.

The basis for these difficulties has been described by Ravetz (1980). He points out that the "cognitive aspects of thinking about risks" may be summarized by the fact that they are universal, incalcuable, and conceptually uncontrollable. (See also the discussion in Chapter 5.) Almost all the decisions made by an individual are viewed by that person in terms of his overall experience. This is also true of his reaction to technological hazards, especially when they include the decision to undertake some action. Thus, whenever an analysis attempts to include reference to the impact of some event or condition on the affected individuals, the analyst confronts an impossible situation. A complete description of the impact on a set of individuals would have in refer to the total individual experience of each person involved. Hence, any analysis must be incomplete.

Further, since the occurrence of accidents is generally difficult to anticipate with any precision, even less complex technological hazards are "incalcuable" in a fundamental sense. More complex hazards, whose effects extend over larger geographical areas, are in some respects "unimaginable". Even the nature of the causal chains that would come into play are sometimes quite speculative. This is even more pronounced when there is no empirical experience with the phenomenon under assessment (e.g., the dispersion, burning or explosion of a cloud on liquified gas escaping from a ruptured storage vessel).

The "conceptual uncontrollability" to which Ravetz refers derives from this uncertainty about the actual course of events and the possible consequences. Even after an accident has occurred, one cannot be certain how much more would have had to be done to prevent its occurrence, or whether that degree of commitment would have been a reasonable precaution.

The impossiblity of comprehensive analysis does not mean that all analysis is without value. To the contrary, the basic contribution of analysis is the selective simplification of events and situations so as to reveal the crucial elements and interactions. The aim should not be to be comprehensive, if by comprehensive is meant the inclusion of every factor influencing each element of every component of the system being examined. Rather the goal should be to include only the relevant factors and important interactions. A great deal of methodological work has been done, both in terms of procedure (e.g., sampling, descriptive statistics, measurement theory) and of interpretation (e.g., development of hypotheses and theories in relevant subjects), which can guide the performance of such analytical simplification.

A second point is that analytical studies and their results often turn out to be weapons used against opponents in technological conflicts. In The Politics of Mistrust, Wildavsky & Tenenbaum argue that the energy policy debate in America cannot be resolved by better information and analysis. (Noll, 1981) The reason, they claim, is that energy policy in America is, and always has been (except during the two World Wars), a polarized political issue. Such polarization results when the political differences among the major parties are rooted in different perceptions about the purpose of the policy in question and the values it is supposed to serve. When there is such difference in opinion concerning the basic aims and purposes to be pursued, a given piece of information can produce conflicting policy conclusions in different groups. Moreover, since such differences stem from differing views of the values expressed in the goals of the program in question, each group is likely to mistrust information generated by others. Even relatively unimportant variations in information can cause heated and largely unproductive debate about the integrity of its sources.

This difficulty is often acknowledged in the design of an analysis by including questions like "Who sponsors the study?" "Who performs it?" or "Who are the users of the results?" Such questions are based in the recognition that sponsors, performers, and users often have an agenda of action and hope that the results of an analytical study will help them achieve their aims.

The consequences of this do not so much undermine rational analysis per se as demonstrate once again the need for care in defining the basic questions and design of the study. For exmple, Wildavsky & Tenenbaum suggest that the energy studies of the 1970s actually contributed to the conflict by examining the wrong questions. They claim that the right question was "How long do we have before we must find substitutes for oil and gas as the primary energy source?" Instead, studies addressed administrative processes designed to balance supply and demand, or decisions about when and where to look for new sources. (Noll, 1981)

4 AN ANALYTICAL FRAMEWORK FOR DECISIONS CONCERNING TECHNOLOGICAL HAZARDS

Previous chapters have described the development of analytical techniques for evaluating what to do about technological hazards. Common approaches and concepts were described, together with at least a brief account of the bases of each approach and its limitations. It should be clear that there are reasonably clear ideas about how to orgnize information for the technical and scientific judgments required, but that there is much less agreement about how to view and contribute to the political judgments which are equally a part of the overall process. The aim of this chapter is to address this latter point, to provide a model of decision making about technological hazards which incorporates each level of societal aggregation--the individual, groups and organizations, national governments and other soverign polities, international organizations and agreements--in a way that enables one to see what kinds of information play a role in different decisions.

But before setting out the elements of the model, perhaps it is good to set the stage with a well-known example. It will help emphasize the political themes which are the focus of this chapter in contrast to the technical and analytical topics that have held the limelight up to now.

The accident--or incident, as it is called by people in the nuclear industry--at Three Mile Island is one of the most widely publicized industrial mishaps in recent years. Robert C. Williams, who together with Philip L. Cantelon wrote the official history of the U.S. Department of Energy response to the nuclear accident at Three Mile Island, has characterized the events there as a failure not so much of technology as of technocracy (i.e., the rule by experts based on criteria of rationality, efficiency, productivity and profit). (Williams, 1980) In describing the events associated with the accident, Williams defines seven levels of concern and human decision making which derived from what happened. He characterizes them as concentric circles, the smallest within the Unit 2 reactor at Three Mile Island, with each larger circle referring to a wider frame of reference.

Williams calls the innermost ring the circle of technology. It encompasses the machines which make up Unit 2. The accident began with a malfunction of these machines. The demineralizer that removes silt from river water entering the reactor system clogged, blocking the supply of secondary cooling water, which caused the reactor to shut itself down. The machine functioned properly, and within seconds activated emergency cooling systems. Unfortunately, humans intervened from the outset, making choices that turned out to be in error. Maintenance men clogged instead of clearing the demineralizer. Earlier, maintenance personnel had closed two valves elsewhere in the system, and that fact was not recognized for some time. This exacerbated the heat problems arising from the lack of secondary coolant.

Then events moved to the second level, the circle of technocracy. The operators in the Unit 2 control room became involved. Within seconds of the malfunction in the machinery room they knew something was wrong. But they did not know exactly what it was. Because the closed valves were overlooked, the operators thought the emergency cooling system was functioning properly. They felt the immediate danger was that too much rather than too little water might enter the reactor core. However, no water was flowing into the core. Nevertheless, in the minds of the operators the problem was not too little water, but too much. Within minutes they made a choice which appeared to be correct, but turned out to be wrong. Further, it was several days before this misconception, and those which followed from it, could be corrected. The training of the operators led them to imagine a situation of too much water in the reactor, and their reactions led them to contribute to a loss-of-coolant accident.

Williams calls the third ring the circle of containment. By this he does not mean the containment building, but the entire plant. Apparently the core materials were successfully contained within the reactor. What escaped into the atmosphere were mainly inert gasses from the radioactive primary loop water. The primary loop fluids usually circulate within the reactor and give up heat generated there to the secondary loop water, which is then cooled in the cooling tower. But during the course of the accident, primary loop fluids flowed out of the reactor onto the floor of the auxiliary building, from which it escaped through vents into the atmosphere. Some of these releases were due to deliberate choice. For several days Metrapolitan Edison officials told the public that releases were nonexistent or accidental (when in fact they were intentional) and that the reactor was under control (when it was not). The key people in the circle of containment were the plant managers, who had a great deal to lose economically as a result of the accident. They were as concerned about containing information about the accident as with containing the radiation. On this level the greatest fears were radiation and news about radiation. The failure of the technocrats at Three Mile Island to communicate clearly and informatively to the public helped generate panic and fear.

The fourth circle lay outside the plant in the emergency response efforts made by local, state and federal officials. Williams calls the fourth level the circle of emergency response. The main problem was to monitor the radiation coming from the plant and then take measures to protect the public from it or from some more catastrophic possibility should one develop. But at this level, information was hard to obtain. There was almost no communication to scientists and officials from within the plant. Helicopter flights were made round the clock in hopes of monitoring releases, but the releases were often unannounced, or announced and then cancelled. In general, the scientists outside did not know what to expect because communications with the plant were so poor. They had no data from within the plant, and did not know how the readings they made were being used by the Nuclear Regulatory Commission or by other officials. They had been led to believe that the radiation releases would be mostly iodine-131, but their readings of that isotope were less than those in the same area following a Chinese nuclear bomb test in 1976. They could not understand why they were seeing government officials on television speculating about evacuating the residents of the area, explosion of the plant, and meltdown of the reactor core.

These first four circles have mostly to do with the technocracy--the unit operators, plant managers and scientists called in for emergency response. The remaining three circles defined by Williams have to do with the public and political reactions to the incident.

Williams calls the fifth level the circle of evacuation. Thousands of citizens had to choose whether or not to leave their homes. Evacuation of people because of floods is common in Southern Pennsylvania, but there were no plans for evacuation in connection with such an accident. As events evolved, there was no evacuation, which implies orderly movement, but rather exodus and flight. People did not wait for the evacuation order that never came (aside from the recommendation that pregnant women and young children leave the immediate area), but simply got in their cars and left. At this level, considerable confusion was generated through imprecise communication. Government officials did not listen to company spokesmen, or heard what they wanted to hear. In response to the question, "Are you ready to evacuate?" a plant manager replied, "We're always ready to evacuate." By this he meant that the plant personnel were always prepared to leave the island if necessary. However, his answer was taken to mean that the people living around the plant should be evacuated from the area. In the end, nobody chose to initiate evacuation of the region. Nevertheless, many people took it upon themselves to seek safety far away.

The sixth circle is the circle of politics. At this level, events involved the media, communications and elected officials. In one sense, a catastrophic nuclear accident had already occurred in the public imagination. _The Prometheus Crisis_ , a fictional account of a nuclear nightmare in California, described a "hundred thousand towering infernos" and "The China Syndrom", a film appearing at the same time as the Three Mile Island accident, characterized a nuclear accident as very hot, very explosive, and so big that it could blow the whole of Southern California sky high. Given such popular expectations the accurate communication of the risks associated with the events at Three Mile Island was crucial. But at first the utility acted as if nothing whatever were wrong, and it was three days before the Nuclear Regulatory Commission sent a team of experts to the site. A single, credible scientific expert capable of communicating what was happening appeared far too late. The fact that scientific evidence indicated low levels of radiation and minimal dangers throughout the course of events was never effectively communicated. Too many decisions were made on the periphery, in Washington, D.C. or Bethesda, Maryland, instead of at the center of events. And the whole thing was described in a language incomprehensible to the lay person.

The final level of concern is termed by Williams the circle of mortality. Subsequent reports of psychological stress suggest that the main consequence of Three Mile Island may be the public's sense that it narrowly escaped death, and that life is out of control. For many people, a nuclear reactor is simply a silent bomb, regardless of the innumerable statements that such reactors cannot possibly explode. The unseen and poorly understood threat of radiation adds to this feeling. It is here, on the seventh circle, that Williams places many people coming in their imaginations to nuclear hell. People see cooling towers, but their subconscious reacts to a fireball and sterility, cancer and stillborn children.

The nuclear accident at Three Mile Island was unquestionably a dramatic example of what can happen when technological hazards go wrong. But it cannot be considered an isolated case. Burton and his colleagues describe a similar situation associated with an evacuation Canadians residents following the derailment of railroad tank cars containing propane gas and chlorine gas (in separate cars) and the resulting explosion and fire. (Burton et.al. 1981) Other such incidents were a chemical explosion at Flixborough in England and a release of an extremely toxic gas at Seveso in Italy.

Such accidents constitute the extreme in the management of technological hazards. They are the emergency that most safety procedures are designed to avoid. But because they are extreme, and because they are subject to public attention, they bring certain points into strong relief. In particular, an emergency like Three Mile Island demonstrates dramatically that the influences on resoprse to technological hazards are by no means restricted to technical factors. To the contrary, the more important elements in an emergency situation are generally human interactions. Poor communications, inappropriate expectations, motivations to protect or advance one's interests--these and other nontechnical factors can dominate the situation. The concepts presented in this chapter are aimed at incorporating both the scientific and technical factors as well as these social and political factors in a common analysis.

Typical Patterns of Interaction with respect to Technological Hazards

The preceding example demonstrates the complexity of the interactions among various groups in the management of a technological incident. Previous chapters have suggested the complexity of the technical issues which are involved in the analysis of such events and the likelihood that they might occur, as well as described approaches for dealing with those complexities. This section presents a simple model which can be used to characterize the interactions among various individuals, groups and organizations with respect either to analysis of or response to an event.

Figure 4.1 shows a schematic diagram of four principal sets of individuals and/or groups or organizations and their typical interactions with each other as well as with the technology presenting the hazard. The four groups are:

> Technocrats (T): those experts involved in the operation and management of the technology in question
>
> Public Officials (O): those who oversee the development and control of the industry employing the technology
>
> Politically Active Individuals (P): those attempting to incluence public opinion and government policy
>
> "Uninvolved" Individuals (I): those exposed to the hazard or suffering its consequences

In the stylized diagram in Figure 4.1, an individual technocrat (represented by the letter "T") is placed within an organiztion or group (represented by the circle around the letter). Each set of individuals is so designated: the letter both represents a

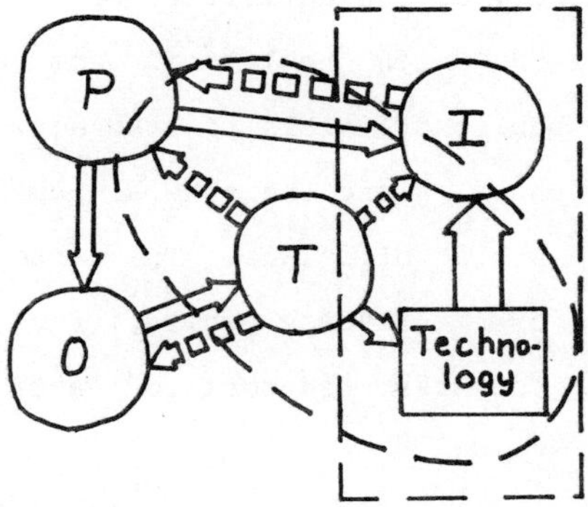

Figure 4.1 Sets of Individuals, Groups and Organizations and their Typical Interactions with respect to Technological Hazards

single individual and designates the whole set. What we usually think of as the industry using the technolgy in question (characterized by the box labeled "technology") is portrayed by the broken elipse encompassing the technology, the set of technocrats, and a part of the "uninvolved" individuals. The sequence source-pathway- consequence-reaction (the basic concept in the identification of hazards as described in Chapter 2) is represented by the box shown by broken lines. Each person belongs to a set of individuals who share his characteristics, and each set, whether it is made up of informal groups or formal organizations, is within the overall context of society.

The solid arrows in Figure 4.1 represent the principal routes of influence among the various sets of individuals and groups. For example, it may be said that the technology is operated by the technocrats, and that the influence of public officials in either stimulating development of a technology or controlling its undesirable consequences is not directly on the technology itself, but rather indirect through the technocrats. Similarly, the majority of the activities of those who respond to a particular hazard with political action are aimed at influencing either public officials to take action or at building a broad base of support among the general public which can then be used to motivate public officials. The dashed lines represent secondary routes of influence which, for the most part, contribute to the long-term support of particular policies or systematic changes.

The membership of a given individual in these sets may change with respect to different issues. The expertise required for controlling nuclear technologies may or may not qualify one as an expert with respect to the technologies of chemical production. They most likely would not be relevant for the newly developed production technologies based on techniques of genetic engineering. It is also typical to find individuals involved in political activities of the type referred to here to be active for only a restricted set of issues. Although individuals who are sufficiently motivated to undertake political action on one issue are likely to be active with respect to other topics as well, it is impossible for one person to respond to all the potential hazards found in contemporary society. Therefore the individuals actually engaged in political action are likely to vary for different issues. Similarly, public officials have authority for specific areas of responsibility, so that different officials are likely to make up the set of public officials for different technologies and their attendant hazards.

Of course, this stylized schematic is considerably simpler than the situation one should expect to find in an actual case. For example, there would be several groups in each set, and each group would exhibit some unique characteristics. However, despite its simplicity, the diagram in Figure 4.1 is sufficient to begin the differentiation required for the characterization of the social and political factors influencing the management of technological hazards. The concepts in subsequent sections will enable the elaboration of these elementary elements and interactions in ways better able to capture the events we are able to observe in actual cases.

Decision Making in a Pluralistic Setting

The next step is to characterize decision making in a setting such as that outlined in the previous section. The emphasis in this will be on the decision to act. Although in some respects it is the action itself which is of ultimate interest, the decision to do something (as well as the choice of what to do) will command our attention. There are several reasons for this. First, it enables us to focus on individual decision makers. This allows the analysis to concentrate on concrete entities, individual persons, rather than constructs like "group" or "organization", which are often difficult to define. Informal groups and formal organizations alike can then be described in terms of the interactions between and among the individual decision makers that make up the group or organization. A large portion of this section is devoted to describing the ways that individual decisions aggregate into collective decisions.

Second, emphasizing the decision to act gives implicit importance to human volition. Lindbloom (1977, pp. 134-135) has described four fundamentally different ways a person can choose: 1) he may simply respond to taste and personal preference, 2) he may make a complex judgment, hastily or after considerable thought, 3) he may choose according to an ethical or moral code, and 4) he may treat as a simple preference, irrationally most would say, what others consider a complex judgment. There are two important attributes of human volition which affect the overall description of decision making in this chapter. It encompasses more than the simple and direct response to conditions referred to in the concept of economic preference. Of course, the concept of utility maximization could be extended so that doing what one ought, however much one would rather not, is in fact a method of maximizing one's utility. But then the concept looses its usefulness to lable a particular kind of motivation. A preference is a fact and can be discovered. But a choice includes volition, an emergent act of will. Only after choices have been made can they be referred to as data. The second attribute of human volition is that the performance of a choice should be thought of as embedded in a process that forms volitions as well as produces actions. Preferences might be thought of as exogenous to the decision making process, formed by conditions and events that may have nothing to do with the question at hand. But to form volitions, the individual needs information and advice. He needs to weigh the similarities and differences of his own thinking with those of his fellows. In matters of preference, there is little need to consider the preferences of others. But when a choice is the expression of volition, questions of fact, moral guidelines, personal desires and interactions with others all need to be considered. Rather than striving for a compromise, one often attempts to integrate different positions through re-forming volitions in the light of the conflicts which have been identified.

A third reason for focussing on the decision to act is that it provides a natural point for distinguishing between the factors which influence an action and the act itself. Of course, the implementation of many complex decisions requires sustained action over time, and it is often necessary to adjust or modify what is done on the basis of what happens. But such subsequent choices can be treated as separate decisions, each with its phase of contemplation and action. However, there is no need to be overly complicated in characterizing the point at which a decision is made. The important point is that a decision to act constitutes the integration of all the various influences into a single conclusion by the individual.

The discussion of typical patterns of interaction in the previous section made clear that decisions to act in response to technological hazards, in so far as they constitute the expression of volition rather than preference, must often be made with reference to complex interactions with other individuals, groups and organizations. The previous section presented a simplified diagram of such interactions which does not incorporate the full complexity one usually observes in actual examples. For example, different levels of government are often involved, so that one must refer to public officials at the local, state and national levels. There is often similar differentiation within the ranks of the technocrats, with single firms and industrial associations on the one hand, and individual experts and researchers belonging to professional associations on the other. The politically active also break down into a variety of groups, ranging from the traditional political parties through citizens action groups (Buergeriniativen) to grass roots groups interested in local problems.

The model discussed in this section attempts to describe decision making in terms of this interaction among many groups and organizations. It describes the individual decision to act in a way which permits reference to these groups and the interaction with and among them. Starting with the individual decision maker, reference to interaction with other individuals or groups and organizations can be expanded to incorporate whatever levels of government is appropriate. Reference to international organizations, or to bilateral or more complicated agreements can also be made if necessary. Table 4.1 lists the basic elements of the model.

The first point in Table 4.1 deals with what might be called the "passive" influence on the decision to act, or those which are exogenous to the process of deciding. Another way of

1. Individual action is subject to three sets of influences or constraints:
   - individual characteristics
   - group or organizational characteristics
   - social context

2. Individual action intended to affect interaction between or among individuals, groups or organizations can take several forms:
   - iniation
   - veto
   - control
   - brokering

3. Group or organizational behavior is an aggregation of individual action, determined by the following:
   1. marshal facts and arguments
   2. secure support of subordinants (if any)
   3. obtain support of colleagues (or peers)
   4. win support of the constituency (of the group or organization)
   5. seek support on a broader basis

4. Support is sought through a combination of:
   - authority (direction of others)
   - exchange (quid pro quo)
   - persuasion (agreement of the need for action)

Table 4.1 Elements of a Model for Decision Making in a Pluralistic Setting

characterizing them would be to say that they represent the setting within which the decision process is played out. With respect to a particular decision they may be thought of as static, although over time they may gradually change.

The importance of individual preferences, moral and ethical rules, and volitions for individual choice has already been described. All these, together with the degree of knowledge of relevant factors as well as other individual characteristics, make up what is sometimes called the "world view" of the individual. Although the way an individual sees the world may change over time, for the purposes of a single decision it may be said to remain constant, and is exogenous to the process of deciding.

Similarly, the individual contemplating action will view himself as a member of certain groups or oganizations. These can be more or less formal in thier structure, with different ways of defining membership and exhibiting various kinds of interactions among their members. Because such groups differ in the resources they command, the pressure exerted on members to conform or contribute to their aims, and the strength of differentiation between members and non-members, they can exert differential influence on the individual decision maker. The individual may also be able to exert influence on the group, thereby adapting its aims to his own purpose or using existing interests to his own advantage. Again, for the purposes of examining a specific decision, the characteristics of relevant groups and organizations may be said to remain constant. Of course, such characteristics change over time, and different groups and organizations may be relevant for the same individual with reference to different decisions. Nevertheless, groups and organizations can constitute important constraints (as well as opportunities) with respect to decisions to act on the part of an individual.

Finally, the social context of the individual may be said to exert influence over his decisions. Although the mechanism whereby this influence is exerted may be to affect his world view, it is included as a separate influence. This is because there are a number of attributes of social context that are under-emphasized when describing their impact on the individual's world view which are nevertheless important when contemplating action. For example, organizational patterns of jurisdiction and empowerment, resource and motivation, and accountability are important considerations which probably have little direct influence on the world view of an individual yet are important factors in his social context. Although the totality of his experience is within this social context (which constitutes the greatest determinant of his world view), there are elements of the social context which affect the individual only indirectly, but still remain important influences on decision making. Again, these are factors that change over time, but which may be considered constant with respect to any given decision.

The second point in Table 4.1 introduces the idea that there are several different forms that action can take. The simple notion of the "decision to act" brings to mind some physical act intended to accomplish a particular aim. But a more refined concept is required in order to capture the complexity of decisions concerningf technological hazards. Among the many ways of exerting influence, Cox & Johnson (1973) have emphasized the four in point 2 of Table 4.1.

In the characterization of the accident at Three Mile

Island recounted at the beginning of this chapter, the operators of Unit 2 initiated certain actions as a result of the readings on their instruments. This is a good example of what is referred to here as "initiation". But this term also refers to a variety of acts other than the direct physical intervention in a sequence of cause and effect. Initiation also refers to any action intended to accomplish the particular aims in mind which does so through an increase in the level of some relevant activity. Thus, the establishment of regulatory mechanisms, or of requirements for testing of chemical substances prior to their being brought onto the market, or of mandatory use of seatbelts in autos are all examples which fall under the rubric of initiation.

At a later point in the events at Three Mile Island the plant managers hindered the flow of information from the technicians in the control room to the scientists outside. This is an excellent example of a veto action: an act intended to accomplish the aims in mind (in this case, the protection of the economic position and reputation of the utility) through the elimination of the the relevant activity, its reduction, or the maintenance of the status quo. Individuals and groups attempt to exercise a veto in various ways. For example, citizen occupation of a building site so as to block construction of a power plant or motorway, and motions in the courts to order that activities "cease and desist" are both attempts to exercise veto power. In general, a veto is any attempt to prevent action or to interfere with its implementation. In this sense it reduces rather than increases the level of relevant activity.

The exercise of control derives from a position of authority, and generally involves the allocation of relevant resources. Those who control budgets and assign discretionary funds have the ability to exercise considerable influence over the actions of others. Control need not, however, derive only from the direct allocation of financial and material resources. Those who determine the content of television newscasts, for example, exert a great deal of control over the degree to which particular events capture the attention of the general public. However, whether it deals with the allocation of traditional resources like finances and materials or other resources like access to the flow of information, the exercise of control in the terms used here refers to the allocation among areas which are competing for attention or resources.

The final kind of action in Table 4.1, brokering, refers to the attempt to bridge the gap between factions or positions. It has already been mentioned that people sometimes find it necessary to re-form their volitions in order to resolve conflicts. This process is often made easier when there is an intermediary that can convey the positions of the various factions without the emotions stimulated by a direct confrontation of the different groups. The usefulness of such brokering has become obvious in international affairs, where "shuttle diplomacy" now appears to be a standard option for dealing with crisis. But brokers can be useful whenever various factions hold what appear to be intransigent positions and are unable to resolve their differences, or cannot come into direct disussion with each other. The function of the broker is to seek out the factions at interest and serve as a channel of communication whereby compromise or integration (re-forming of volitions) can be achieved.

Points 3 and 4 in Table 4.1 deal with strategy. The third

point postulates an approach in which the individual shifts his attention according to a particular sequence, thereby enabling both the technical aspects of a particular decision and the potential interactions at various levels to be included. Although orginally developed to describe bureaucratic decisions when public opinion matters (Coppock, 1977), the basic approach can be applied to most settings in which the individual functions as a member of a group or organization. In this study it is assumed that individual action alone is insufficient to constitute adequate response to technological hazads. Therefore, appropriate responses will almost always involve some sort of coordinated action among one or more groups. The sequence in the third point is a simple guide for sorting through the potential interactions.

The first step in the sequence is to marshal the facts and arguments upon which the individual's position is based. This refers both to the development of the concepts and to the checking of facts and figures. It should insure the correctness of the technical, scientific and economic bases for the action contemplated. Procedures which might be used for this task have been described in Chapter 2, especially under the headings "Safety and Reliability Engineering" and "Cost-Benefit Approaches". It will be recalled from those descriptions that the process can be quite difficult. In some cases there will remain unanswered questions, leaving the correctness of the conclusions open to challange. But the point is that the soundness of the contemplated action should be examined before trying to build support for it. Of course, there are situations in which an individual can build support for his position regardless of its contents. Doing so, however, runs counter to the overall aim of this study: to be rational and intentional in dealing with technological hazards. It will always be assumed in this study that the initial step is to develop and check a sound technical, scientific and economic basis for any action being contemplated.

Once the "argument" is in order, the individual turns to developing a basis of support for his action. In this process, he begins with those associates with whom he has greatest influence, and gradually shifts his frame of reference to include larger numbers of people. Thus, an individual in an organization would typically begin by securing the support of his subordinates, if anyone reports to him. He then shifts his attention to those of equal status with himself, his colleagues or peers. Obtaining adequate support from one's colleagues can be difficult, especially if there are several factions or if one is in direct competition with the others. Next, one has the option of attempting to obtain the support of the top authority in the organization, and thereby the "official" backing of the group or organization as a whole, or of appealing directly to the constituency of the group or organization. ("Constituency" is used here to refer to those individuals or groups with direct exchange of information, funds or materials with the organizition in question. It is that portion of the population relevant to the organization as an "open system"--see the discussion in Chapter 6.) It is not always necessary to obtain the official support of one's organization to undertake action. In informal groups, and in organizations which are less "authoritarian", it is not uncommon to find individuals who regularly rely on support from outside. Finally, after support has been sought in the immediate constituency of the group or organization, attention can be given to obtaining support on a broader basis. This can be done by concentrating on specific groups or organizations, by broad appeal to the general public, or by a combination of both.

Of course, the sequence as outlined here is only a guide. It is a simple procedure which enables one to refer to the technical aspects of an issue as well as to the interactions to be expected at various levels of social aggregation. It must be combined with ideas about how to win such support before it becomes practicable.

Point 4 presents the three fundamental ways one can obtain support for his position. Lindbloom (1977, p.4) argues that there are only three basic mechanisms for social organization: "One is social organization through the authority of government. One is social organization through exchange and markets. Another ... is social organization through persuasion...." He goes on to show that all the various socio-political systems in the world can be described in terms of different combinations of these three fundamental mechanisms.

Lindbloom describes authority as the basic relationship that characterizes membership in formal organizations. When one says that he belongs to an organization, he recognizes the authority of its officers to act for him in certain matters or to dictate his actions with respect to the function of the organization. Whether this authority is obtained through the "legitimate" agreement of the members or through coercion, once it has been established any single application of authority is costless: the marginal cost of control is nearly zero. Grants of authority can be made long in advance, and can embrace large categories of acts. When response is needed, no reward, no punishment, no manipulation is required. All that is needed is the indication of the response desired. (Lindbloom, 1977, pp. 17-20) But authority must be maintained. Over the long run, authority requires willing participation by those subjected to authority, which enables the "subjects" to exert a degree of influence over those who exercise authority. Even when the majority of the population is coerced into accepting authority, there is a smaller group who enforces that authority. The higher officials must then insure the cooperation of this smaller group, and are thus open to influence. In some circumstances officials grant authority to subordinates to obtain their cooperation as well as to organize the social system. This suggests two aspects of authority that an individual might make use of to advance his aims. First, the simple and direct exercise of authority over his subordinates, and second, influencing his superiors through their need for his cooperation.

Lindbloom (1977, p. 33) claims that just as hierarchical, bureaucratic and governmental systems arise from the authority relation, so market systems arise from the exchange relation. The exchange relation, he says, is one of deliberate control. Two (or more) persons offer something of benefit in order to induce a response, and the offers are contingent upon achieving the desired response. Thus, exchange is not merely a method of reshuffling posessions, but also a method of controlling behavior. A certain sort of exchange also plays an important role in organizations structured according to the authority relation. It is often possible to exercise authority in such a way that one not only reinforces his own authority, but so that benefit also accrues to someone else. Thus, it is possible to negotiate a kind of exchange within authority structures, whereby an accomodation of interests is achieved. It is not a true exchange of authority, but more like an exchange of favors: the commitment to exercise authority in a particular way in exchange for the desired response. Such accomodations should be distinguished from other methods of control in authority structures. For example, sometimes authority grants are made with the expectation that the desired response will occur, but without the quid pro quo of exchange. (A reduction of taxes, for example, is a grant of a benefit which may produce the desired response, but which is not conditional upon it.) This suggests three additional ways an individual might obtain support for his position: outright exchange of goods or services, accomodation in the exercise of authority, or unconditional grants of benefit.

The third mechanism of social organization is persuasion. Lindbloom (1977, p.13) points out that persuasion is central in all systems:

> In the form of ideological instruction and propaganda, persuasion is a major method of elite control of the masses, much more so in communist systems than in liberal democratic ones. In the form of commercial advertising, it is a major instrument of corporate control of masses of consumers in market societies. In the form of mutual persuasion in "free" societies--that is, in the form of "free competition of ideas"--it is fundamental in liberal democracy.

It is this latter aspect of persuasion which is of greatest interest here, although the other two should not be forgotten. In particular, the approach of "education" in what Lindbloom calls preceptoral systems (social control based on unilateral persuasion) is instructive of how persuasion works. In a preceptoral system (examples include the older fascist and nazi regimes as well as those currently in mainland China and Cuba), rationality rests on an ideology which "once taught to the individual gives him both a 'correct' understanding of the social world and guidelines for his own decisions". (Lindbloom, 1977, p. 59) Although preceptoral systems rely on moral and emotional appeals to supplement the rational, Lindbloom claims that at their core is a genuine attempt to raise the level of a person's conscious, thoughtful, deliberated understanding. This is equally the core of persuasion as a method of obtaining support for response to technological hazards. The individual encourages others to recognize the need for action. He calls upon them to acknowledge the common problem and appeals to personal pride and sense of responsibility as motivations for action. But this appeal must in the end be based on convincing others through the plausibility of his argument. Of course, even questionable content, repeated often enough from a variety of sources, sometimes becomes convincing. That is one of the techniques of preceptoral systems. But using persuasion to gain support for rational and intentional response to technological hazards in the spirit of this study should be based on the plausibility and convincingness of the argument.

A Conceptual Framework for Managing Technological Hazards

Based on the foregoing, a conceptual framework for the management of technological hazards can be developed. It is not a true model, because it does not speicify the interactions among its various elements. It merely lists the different topics that might come to the fore in in various technological issues. But it sets them out in such a way that both the technical and scientific judgments can be seen in clear perspective. Taken together with the other topics presented in this chapter--typical patterns of interaction with respect to technological hazards, and a guide for decision making in a pluralistic setting--this conceptual framework can provide a basis for the analysis of technological hazards and possible responses to them.

In simplest terms, such analysis requires activity of three quite different kinds. First, evidence relevant to the technical and scientific as well as to the political judgments involved needs to be collected and placed in proper perspective. Second, information concerning the opinions, volitions, and likely actions of relevant interest groups should be gathered. Finally, a program of response should be designed and implemented. However, in the conduct of an analysis, these three activities cannot really be separated. There is a high degree of overlap between the political judgments required in evaluating the importance of the consequences of a particular hazard and the anticipation of how various groups might respond. Similarly, there is considerable commonality bewteen the tasks of identifying a hazard and its consequences and those of designing intervention measures aimed at reducing its occurrence or the impact of its consequences.

It is proposed here that the analysis of technological hazards be organized around four tasks: 1) the specification of causal relations, 2) the characterization of priorities, 3) the assessment of response capacity, and 4) the design and implementation of intervention measures. These four tasks clearly reflect greater emphasis on topics which are likely to contribute to on-going response to a hazard--monitoring and permanent intervention--than has been the case in previous schemes. (See chapter 2.) Not only is greater attention given to the kind of information required for the establishment and maintenance of intervention programs, but the political and conflictual nature of the effort is clearly recognized. Both scientific or technical judgment and political judgment are required in each of the four tasks, and such judgments will likely have to be made on the basis of incomplete or uncertain knowledge, which will often result in alternative and conflicting interpretations. Recognition of this is essential for a rational and intentional treatment of technological hazards and response to them.

The complete conceptual framework is present in Table 4.2. Each task is broken into subcategories which describe sets of information and foci which contribute to the organization of the overall task.

The specification of causal relations incorporates two sub-activities: identification of the hazard and its consequences, and estimation of the likelihood that various events in the sequence from the originating event to the ultimate consequence might occur. Certain techniques useful in these tasks, especially fault trees and event trees, have already been

Specification of Causal Relations
   Identification of hazards and consequences
   Estimation of the likelihood of occurrence

Characterization of Priorities
   Assessment of awareness and conflict
   Evaluation of the consequences

Assessment of Response Capacity
   Organizational capacity
   Professional capability

Design and Implementation of Intervention Measures
   Specification of the scope
   Selection of the structure for management and control
   Implementation

Table 4.2 A conceptual Framework for the Management of Technological Hazards

described in Chapter 3. But other approaches are also useful, particularly in the identification of hazards in the first place. Kates (1978, pp. 14-19) claims that four methods constitute the principal ways in which the existence of hazards are demonstrated: research, screening, monitoring and diagnosis. He dismisses the identification of hazards through pure, or basic, research as "haphazard", and claims that: "The institutionalized task falls on 'practical' or 'applied' science for the identification of hazards by screening, monitoring or diagnosis." (Kates, 1978, p. 15) He defines screening as the application of a standardized procedure to classify products, processes, phenomena or persons; monitoring as recurrent observation, recording and analysis of products, processes, phenomena or persons; and diagnosis as the assessment of symptoms or consequences in relation to possible causes.

Burton & Whyte (1980) employ a somewhat different scheme to describe the identification of hazards. They divide the activities into three categories: modelling, monitoring and surveillance, and screening and testing. They describe modelling as ranging from simple characterizations of processes and interactions in flow diagrams to complicated computer simulations. Monitoring is defined as the process of repetitive observation, for defined purposes, of one or more elements in space and time, using comparable methods of sensing and data collection. Monitoring can be undertaken for different puropses: to understand the processes involved, to provide evidence of damage, to contribute to managment and control decisions. Testing is defined as the use of standardized procedures for measuring sources or undesired effects. The characterization by Burton & Whyte emphasizes that one need not use the most complicated methods available to be rational in the examination of technological hazards. A well thought out flow diagram of the phenomenon in question may produce more useful information than the attempt to provide the most precise measurements possible of only those parts of the causal chain that can be measured easily.

Similarly, there is a broad range of techniques, exhibiting various levels of sophistication and precision, available for the estimation of the liklihood that potential events will occur. These include: preparation of taxonomies, analysis of logic trees (event trees, fault trees and decision trees), writing of scenarios, and simulations (either physical or mathematical models).

A taxonomy, or an orderly classification of the potential sources, pathways and effects, is in one sense the beginning of any assessment of a technological hazard. Despite the limited capacity of a taxonomy to reveal the complexity of interactions among the elements it classifies, such a simple ordering can in some situations be combined with scientific judgment to provide a reasonably accurate impression of the phenomenon being examined. When very little data is avaiable, the cost of doing more might be prohibitive.

The next step is to develop a schematic representation of the interactions exhibited by the phenomena being studied. Different levels of complexity and of quantification can be portrayed in schematic flow diagrams. The general procedure is first to develop a non-quantified model which maps out sources and pathways. Numerical representations of the flows or levels at various points in the diagram are then developed. The simplification of a flow diagram is both its strength and its weakness. While a flow diagram can be clear and easily understood, it may misrepresent what actually happens. Nevertheless, flow diagrams can be used to clarify which items are most important, an important definite over a simple classification.

The use of logic trees, particularly of fault trees and event trees, has been described in detail in Chapter 3. The introduction of these techniques is widely acknowledged to have increased the sophistication and precision of the analysis of complex technologies. However, applications of these techniques are also open to challange. For example, the U.S. Reacor Safety Study (also called the Rasmussen Report or WASH 1400) has been severely criticized despite the considerable effort invested in the sophisticated use of quantiative techniques (especially fault trees and event trees). Among the objections raised have been: 1) the system examined is simply much too complicated to quantify, 2) the data base on component failure rates does not support the calculations of the study, 3) improper and incorrect statistical procedures were followed, 4) human behavior was inadequately treated, and 7) the role of quality assurance and failures thereof were inadequately treated. (Lewis, 1980, p. 39) Despite these criticisms, the Rasmussen Report is considered a substantial improvement over previous efforts. It is interesting to note that even after the accident at Three Mile Island, it is impossible to say whether the estimates of the likelihood of core meltdown in the report are accurate. However, the report did note, as have other studies, that regulatory procedures did not give adequate attention to temporary abnormal conditions (called transcients), small loss-of-coolant accidents, and human error. These were exactly the three main ingredients in the Three Mile Island incident: a transcient (water loss), exacerbated by human behavior (the blocked valves), followed by a small loss-of-coolant accident and subsequent human error. Thus, one may conclude that whereas logic trees constitute a powerful and sophisticated tool, their application is neither simple nor straightforward.

Scenarios are used to describe possible events in society as a whole or in a system too complex to describe in quantitative terms. Porter et.al. (1980, p. 150) describe a scenario as a descriptive sketch or chronological outline of a possible future state of society. A scenario is to present a holistic view of the pertinent social context as it relates to the hazard being assessed. They point out that two types of scenarios can be differentiated. One describes a future state of affairs, a "snapshot" of possible developments. The other presents a "future history", a plausible chain of events leading from the present to a particular future state. Scenarios are most useful when they are contrasted with other scenarios, and most effective when the focus on significant differences. However, because they are usually applied to systems too complex to quantify, they are usually quite speculative and therefore present only quite tentative conclusions.

A given phenomenon can be simulated either with a physical model or with a mathematical representation of that phenomenon. Examples of physical models include the scale models used in wind tunnels or other laboratory tests, hydrologic models used to study the collapse of dams, aircraft and vehicle simulators used to test and train operators (sometimes combined with mathematical models through computer interactions), and the practice towers used to train firemen under realistic conditions. A mathematical simulation model constitutes an attempt to express the essence of the phenomenon being studied in mathematical relations. The key factors are given quantitative representations, and the interactions among these factors are described by mathematical equations. Deterministic simulation models use explicit, point estimates of the variables in the equations in the model and the external conditions which influence it. Such models are deterministic in that the result is always the same for a given set of variables at the start of the calculations. Probabilistic simulation models, in contrast, characterize the factor values as scattered randomly about some average value. Thus, repeated calculations using a probabilistic model and the "same" input values will produce somewhat different results. Simulation with a probabilistic model is repeated until the results converge to a single outcome, which is taken as the most likely way the phenomenon will develop. It is not always possible to model a phenomenon using a scaled-down physical replica, and it is always difficult to know if one has really captured the significant factors in a mathematical model. Thus, although simulation models can provide a great deal of useful information, they also can prove to be irrelevant to the actual event being studied.

The characterization of priorities is also split into two sub-categories in Figure 4.2: the assessment of awareness and conflict, and the evaluation of the consequences. The activity of evaluating the consequences of exposure to a hazard has traditionally been viewed as a part of the risk analysis process (see Chapter 2). This has usually been combined with the identification of hazards and consequences, and the estimation of likelihood of occurrence to make up the "package" of risk analysis tasks. However, the approach proposed in this study gives greater emphasis to the "political" elements, which requires both that consideration of the positions and volitions of relevant groups be included in the analysis and that the problems of institutional organization and coordination be included in the design and implementation of intervention measures. The aim is not only to base the choices on values (hopefully) held widely in the society, but also to refer to the volitions of key individuals and groups.

There are three general approaches that can be used in assessing awareness and conflict: interest group analysis, survey approaches, and general surveillance. Interest group analysis is quite similar to the preparation of case studies, and can rely on historical documents (e.g., published surveys, governmental statistical materials, publications, brochures or newspaper articles) and interviews with the principals. It is important that the relevant groups are identified prior to data gathering so that proper emphasis can be given to each relevant group. Otherwise, there is a danger of biasing the view of interest groups through the data base.

Surveys can be formal (carefully structured) or informal. The logistics of of a survey vary with the form, the information sought (e.g., to determine what people do, what people are, what people believe, or what people prefer), and the selection of participants. Some of the issues considered in preparing a survey include: the definition of a representative sample, the construction of an appropriate questionnaire, assuring the accuracy of the answers, and the manpower and other resources necessary to implement the survey. (Porter, et.al. 1980, pp. 124-125) Adequate survey prodedures are not easy to develop. The method depends on a delicate process of communication in which slight differences in wording of questions can dramatically alter responses. However, survey methods have been extensively developed, and the approach can produce useful information about large numbers of people when properly used.

General surveillance refers to the attempt to recognize potential areas of conflict that might have been overlooked elsewhere. It can begin by the examination of four key sets of possible social impacts: households (including individuals living alone), communities, organizations and societal institutions. Data on the following topics should be examined: population density patterns, population mobility, housing, displacement of people (in terms of career patterns as well as location), transportation, community growth or decline, institutional dynamics, health, noise, leisure and recreational activities, and aesthetics. A broad examination of such indicators of potential areas of social concern may identify conflicts that otherwise might be overlooked.

The evaluation of consequences is, along side the specification of causal relations, the second area of major emphasis in the analysis of technological hazards. Four different approaches have been proposed for the evaluation of the consequences of exposure to technoogical hazards: direct comparison with exposure to other hazards, maximization of utility through comparison of risks and benefits, comparison of the relative costs of making a wrong decision, and adjudication.

In an early contribution to the methodology of risk analysis, Starr (1969, p. 1237) observed that the American public appeared willing to accept "voluntarly" risks roughly 1,000 times greater than they accepted for those "involuntarily" imposed upon them, and that the statistical probability of death from disease appeared to be a rough psychological yardstick for establishing the acceptability of hazards. Since then, the comparison of the risk associated with exposure to a given hazard with those associated with established (and therefore presumably accepted) hazards has been a standard part of the discussion about technological hazards. For example, the risks associated with the accident at Three Mile Island have been described as: "a minor loss-of-coolant accident that caused them (the public) less radiation than their dentist, less risk than their automobile, and less heart trouble than Edward Teller (known as father of the hydrogen bomb) gets listening to Jane Fonda (a renown nuclear pacifist)". (Williams, 1980, p. 56) As has been discussed in Chapter 2, comparison with other hazards is called the revealed preference approach because it refers to the manifest choices made in similar situations in the past. It assumes that, through a process of iteration, an optimal balance between the risks and benefits of technological activities has somehow been achieved. The approach also assumes that past behavior and preferences are good predictors of the future. Both of these assumptions appear open to question. (Otway, 1980) Direct comparison of hazards, while useful in establishing baseline comparisons, is probably inadequate as the sole basis for the evaluation of a given technological hazard.

The idea of maximizing utility by comparing the risks with the benefits is one of the most intuitive and widely applied approaches to comparison available. It has been described in detail in Chapter 3. The literature on cost-benefit analysis is interesting in one respect, namely that a typical cost-benefit study will include a discussion of the limitations of the approach (sometimes even recognizing the impossibility of monitarizing all of the relevant attributes), but then go on to apply the approach and adopt the conclusions. This will no doubt characterize applications of the approach in the future, as well. The idea of toting up and comparing the advantages and disadvantages of a contemplated action is too intuitively appealing for the formal difficulties to prevent its use.

One approach to utility maximization not mentioned in the survey of literature in Chapter 3 is that of decision analysis. For some reason, decision theoretic approaches have not found widespread use among engineers and technical scientists, so that the approach has not played a very large role in the development of risk analysis up to now. This may be because the technique employs subjective probabilities, a concept which might be disconcerting to individuals with training in the natural sciences. The "subjectivist" school of probability (DeFinetti, 1937; Savage, 1954; Edwards et.al. 1968) treats probabilities as beliefs about an event or proposition whose truth has not yet been ascertained. Thus, the subjectivists claim that there is no "correct" probability to be measured somewhere in reality. (Otway & von Winterfeldt, 1982, p. 251) Although it is sometimes possible to verify the truth or falsity of a proposition to which a probability was once attached, such as yesterday's forecast of today's weather, the rarity of this possibility makes it difficult to confirm probabilistic statements about technological hazards.

Decision analysis most often takes the form of a decision aid for an identifiable decision maker or body. It can be divided into four steps: structuring the problem, formulating inference and preference models, eliciting probabilities and utilities, and exploring numerical model results. (von Winterfeldt, 1980, p. 71) The task of structuring the problem makes use of many of the concepts already described here: hierarchical trees, taxonomies, schematic diagrams. The formulation of inference and preference models generally makes use of expected utility theory (see Chapter 5), often employing a "multiple-attribute utility theory" which enables choices to be formulated in terms of several, sometimes conflicting, factors. The analyst then interacts with the decision maker or body to determine the preferences to be reflected in the analysis. This often takes the form of a value tree summarizing the clustering of preferences at different levels. All this is then used as input to a mathematical model which draws logical conclusions based on this information.

One approach to the evaluation of technological hazards which has been proposed (Page, 1978) might be though of as a "mirror image" of utility maximization. It is the attempt to minimize disutility, and consists of the comparison of the costs of wrong decisions. Two kinds of wrong decisions can be identified. For example, in attempting to demonstrate that a particular chemical substance is toxic, the analysis may incorrectly conclude that a non-toxic substance is toxic, or that a toxic substance is non-toxic (inability to reach a conclusion will be ignored in this brief discussion). Since the test is of the toxicity of the substance, a finding of toxicity is called a positive test conclusion, of non-toxiticity a negative test conclusion. Thus, concluding that a substance is toxic when in fact it is non-toxic is referred to as a false positive finding; that a substance is non-toxic when in fact it is toxic, a false negative finding. The cost of a false positive finding--concluding that a substance is benign when it is not--can be much higher than the cost of a false negative finding. In the case of a false negative, the cost is merely the foregone benefit that could be derived from the marketing of that substance. In most cases, this is a lost economic benefit to the manufacturer. Only in rare cases does it involve a substance with large benefits to its users and for which there is no effective substitute. The costs of a false positive can be catastrophic in comparison. Recently, the Mansville Corporation in the United States filed for reorganization under Chapter 11 of the bankrupcy laws because of suits filed against it by former employees claiming compensation for health effects deriving from manufacture of asbestos. The company currently faces 16,500 such lawsuits. (Mansville Corporation, 1982) Company experts estimate that as many as twice that number may eventually be eligible to file. Since settlements have been averaging about $32,000, the company faces a dilemma. Although in good financial condition today, if the company were to have to settle in every case it considers possible, the financial burden would be far greater than the company could bear. Mansville's reasoning for filing bankrupcy is that when justifiable claims against a company exceed the firm's ability to pay, the assets are liquidated and distributed fairly among the claimants. But since the health effects emerge first twenty to thirty years after the period of exposure, many of the claimants have not yet been identified. In hearing individual health claims, the judge has no obligation to consider the impact on the firm of all claims. The only way that the interests of those as yet unidentified claimants can be protected, according to Mansville, is through the bankrupcy procedure, which requires consideration of all kinds of debts, including claims for compensation as well as any others. The case also involves a second legal issue. The insurance companies have disputed the responsility for coverage, disagreeing over whether the insurance company covering the individual when he was exposed or when the disease emerged is responsible. (Business Week, 1982) This example illustrates amply that the costs associated with treating a toxic substance as if it were benign can be large indeed. In cases where the potential losses are catastrophically large but the likelihood of their occurrence very small, it might be better to compare the costs of wrong decisions (which would be very large for a false positive result) rather than simply balancing likely costs with likely benefits (which would most certainly would be more favorable towards undertaking the activity).

The final approach to evaluation is adjudication. Adjudication constitutes a formal procedure for resolving conflicts. It functions best when there are two sides to an issue, each of which can be called upon to present its case. The guiding principle of criminal law is to limit false positive determinations. ("Better to free ten guilty men than to falsely imprison one innocent man.") This is because there is no adequate compensation for denial of liberty. Once a mistake has been made and a person unjustly incarcerated, there is no way to compensate for lost liberty. This asymmetry in the consequences of being wrong is one of the hallmarks of the criminal justice, and carries over into the civil law system as well. (Page, 1978, p. 233) The rigorous rules of evidence in adjudicatory procedure, and decision rules such as "beyond a reasonable doubt", contribute towards limitation of false positive findings. Of course, the tribunal is not prohibited from considering the consequences of false negative findings in its deliberations. But the procedures and traditions of adjudication contribute a bias towards the avoidance of false positive findings, a bias that the previous paragraph suggests is not altogether inappropriate for certain kinds of choices.

The assessment of response capacity is divided into two sub-tasks: the determination of existing organizational capacity, and the assessment of professional capability. "Capacity" is used here to refer to the resources which can, in the relatively short term, be brought to bear on the problem in question. "Capability" refers to the intellectual and analytical ability to deal with the issues involved. A secondary aspect of capability is the number of individuals with the expertise required, or who are in training.

Three sets of information are useful in determining the extent of organizational capacity for dealing with a particular technological hazard: the applicability of various organizations, their preparedness to respond, and their accountability. The applicability of organizations can be summarized by the terms "jurisdiction" and "empowerment". Jurisdiction refers to the limits within which authority may be exercised. Such limits can be geographical or topical. Typical geographical limits to jurisdiction are those which follow political boundaries. Authority is often restricted, for example, to the geographical area controlled by the political entity in which the enabling legislations was enacted. Topical limits to jurisdiction deal with some aspect of the objects or behavior which is to be controlled. For example, the determinations of the U.S. Federal Drug Administration cannot be applied to most food additives. But such limits on jurisdiction can also refer to the people or groups involved. For example, criminal law courts generally do not have jurisdiction over juveniles, who are accorded a separate set of rights and responsibilities and separate proceedings in special courts. Thus, jurisdiction refers to the assignment of legal authority with respect to the issues being examined and to the location in which the events occur and response is made. Empowerment deals with a slightly different aspect of legal responsibility--the degree to which the legal entity has the right to impose authority on others. Together, jurisdiction and empowerment characterize the legal right and responsibility to respond.

The second set of information needed in assessing the capacity to respond deals with the resources and motivations of those organizations with the right to do so. In particular, there are three areas of special importance: the allocation of resources, the typical patterns of individual career advancement and of organizational development, and the stability of personnel and policies. Budget allocations are specialized, historical, fragmented, repetitive and sequential. (Wildavsky, 1979) Issues are seldom resolved "once and for all" in budgetary processes; there is usually the chance to review the situation in a subsequent budgetary cycle. Although the budgetary process is often full of conflict, because choices must be made among alternative allocations of available resources, it is very unusual for the process to go back on itself and alter allocations already made. Thus, budgeted resources are a strong influence on the capacity of an organization to respond to technological hazards. Another influence, however, is the pattern of reward and career development. If a particular set of behaviors is singled out as the basis for monetary compensation or career advancement, this behavior is likely to receive preferential treatment within that organization. For example, rewarding managers in a firm on the basis of quarterly profits by division (or cost center) will encourage managers to think in the short term and about the performance of the activities over which they have direct control rather than in the longer term development of the firm as a whole. Similarly, the stability of the organization in terms of personnel and policy positions also influences the preparedness to act. Both continuity of policy and of personnel are important for the implementation of programs over time. Frequent changes of policy, or of the top authorities of an organization, detract from the commitment to act on specific issues.

The accountability of an organization is an important factor in assessing its capacity to respond. Like jurisdiction and empowerment, it brings into sharp relief the differences between public and private organizations. In research on social influences among individuals, one of the more consistent findings is that those behaviors which are open to public view are more subject to social influence than are those which are private. Pfeffer & Salancik (1978, pp. 104-106) refer to this in their examination of external controls on organizations. They claim that the visibility of the behavior of an organization and the outcomes of its activities is one of the most important things which holds an organization accountable. In fact, they point out that it need not even be visible: behavior is often inferred from the presence of certain outcomes. But when neither the behavior nor the effects of an organization's activities are visible, it is no longer publicly accountable. This can be a strong influence on the way it behaves.

The assessment of professional capability is in many ways similar to an examination of manpower development and training in relevant specialisms. For example, the European Program on Environmental Health Aspects of the Control of Chemicals, being carried out by the World Health Organization (WHO) Regional Office for Europe, includes an effort to identify the occupational skills required for the fulfillment of this program and to estimate the manpower available (including training programs) in Europe. (Waddington, 1982) This program is attempting not only to describe the analytical tasks and skills required, but also to assess both the existing facilities and manpower available as well as relevant training programs.

In Table 4.2 the design and implementation of intervention measures has been divided into three sub-tasks: specification of the scope, selection of a structure for management and control, and implementation per se. Of course, intervention measures need to make use of all the information developed under the other tasks, and in a sense would be the accumulation of what had been learned in them. Therefore, attention in the design and implementation task focusses on factors associated with performance rather than content.

The first step is to determine the nature and scope of the effort. The focus should be defined as well as the breadth and depth of analysis indicated by the availability of resources. An important aspect of the determination of the scope is the specification of the geographical limits and organizational resources which are to be included. In many respects, the administrative scope of an intervention program will depend upon the resources and skills of its managers in obtaining support for the program. Aspects of this have already been described earlier in this chapter. This specification of scope will likely be an important influence on the set of alterantive responses considered in an intervention program.

The selection of a structure for management and control of the effort is dependent on many factors. The legal and administrative authority upon which the program is based may specify clear responsibilities and reporting procedures. However, the nature of technological hazards is such that it would often be an advantage to incorporate a much wider set of resources than could possibly be assembled in a single organization. There are a number of different structures within which coordination among cooperating organizations might be achieved. These include development of a common framework to which independent activities can be related, establishment of a joint facility charged with making independent activities commensurate, jointly determining a common policy giving coherence to separate activities, central direction of separate activities, and welding of activity into a single agency program. (See Chapter 6 for a more detailed discussion of these strategies.)

In so far as it differs from other items already discussed, implementation deals with the assurance of satisfactory performance. Such assurance is usually achieved through monitoring the progress of the effort, and checking this against the aims defined at the outset (or modified at some later time). Sometimes this is formalized in an evaluation study of the program.

Summary and Conclusions

Although the management of technological hazards does not always involve emergencies, this chapter begins with a description of a technological emergency. The nuclear accident at Three Mile Island is used because it is a well-known incident, and because it clearly demonstrates the importance of non-technical factors in dealing with technological hazards. Nearly any such emergency would show the same conclusion--the railroad tankcar explosion at Missasagua shows a quite simmilar pattern (although more "successful" in terms of communications with the public).

This illustration is then followed by a simple characterization of typical patterns of human interaction with respect to technological hazards. Typically, one finds four sets of individuals, groups and organizations that are relevant for the management of a technological hazard: the technocrats (who operate the technology), the public officials (who influence the behavior of the technocrats and thereby attempt to control the technology), the politically active (who attempt to bring about laws and administrative regulations controlling activities they consider important) and the otherwise uninvolved individuals (who bear the brunt of the exposure to the hazard).

The simple identification of types of individuals, groups and organizations, even together with the observation that each usually exerts influence of a particular kind is not, of course, a sufficient basis for the management of technological hazards. A second necessary component is some sort of guide for decision making. A model of decision making in pluralistic settings is presented which can be used as a guide for deciding how to pursue one's aims under conditions such as those typifying interactions with respect to technological hazards. This model essentially describes a sequence whereby the individual shifts his attention from technical matters to potential interactions at various levels of social aggregation, together with suggestions for possible ways of securing support for contemplated actions.

The final component presented is a conceptual framework for the management of technological hazards. It specifies four main tasks: specification of causal relations, characterization of priorities, assessment of capacity to respond, and design and implementation of intervention measures. This framework is conceived so as to enable not only information concerning relevant technical and scientific judgments to be placed in proper perspective, but also the information required for the political judgments needed to evaluate the consequences and to implement intervention measures.

Taken together, these three sets of ideas--typical patterns of interaction, decision making in a pluralistic setting, and the conceptual framework--constitute what Leibenstein (1977) calls an "analytical framework". He describes the difference between an analytical framework and theory as follows. A theory is a set of relations that is sufficiently specified so that some conceivably falsifiable conclusions may be reached. Thus, a theory says something about the world of facts, for at least some conclusions of theory are of a form for which it is possible that the facts do not concur. Theories enable us to make assertions about the world of events, and under some conditions, make predictions. For example simply to say that supply and demand follow simple relationships--e.g., quantity demanded depends upon price; quantity supplied depends upon price--is not sufficient to constitute a theory because specific events cannot contradict the relationships as defined. However, if the relationship is so specified as to indicate what would happen when there is, say, an increase in price, then the idea is a theory rather than an analytical framework.

The ideas in this chapter are offered in the spirit of an analytical framework and not as a theory. The intention is to develop a set of ideas with which one can develop a rational and intentional response to technological hazards. Whenever possible, illustrations are used which are drawn from actual experience with technological hazards. But there is no intention to develop a set of concepts for proof against actual events. It is unlikely that any response to technological hazards has employed the approach proposed here. However, this fact could not be considered a refutation of these ideas. Rather the test should be their usefullness in providing an organizing structure for thinking about response to technological hazards. The aim of this study is to suggest how information for the decision about how to respond to technological hazards ought to be organized, not how such decisions are made today.

5 REACTION TO TECHNOLOGIES AND THEIR HAZARDS

Public opinion polls (e.g., Commission of the European Communities, 1979), survey research (e.g., Vleck & Stallen, 1980; Renn, 1981a) and social observers (e.g., Tucker, 1977) commonly refer to increased public awareness of technological hazards and their attendant risks. Although it is widely acknowledged that public attitudes and beliefs are central factors in decisions about certain technologies (e.g., nuclear power) and are likely to be of increasing importance for decisions about technology in general, there is considerable disagreement about the appropriateness of those attitudes and beliefs held by the public.

It should not be surprising that the importance of public beliefs should become obvious first with respect to issues like the provision of primary energy (mostly electric power). The strength of public beliefs, together with the willingness of certain groups to respond in ways which disrupt the ordinary course of development planned for the technologies involved, has become a crucial issue today. It is so central to the functioning of the overall economy that this conflict is one of the key factors referred to by those who question whether contemporary society is governable in the traditional sense. Thus, the challange to technological "progress" stands in the center of current affairs.

This chapter addresses the intellectual and emotional reactions to technologies and their hazards, as opposed to behavioral response to them, which is the topic of Chapter 6. No attempt is made here to review the extensive literature dealing with human reaction in general. Rather the aim of this chapter is to present a few relatively simple, coherent and fairly consistent ideas which can serve as a basis for describing reaction to technologies and their hazards. A broader review of the literature may be found in Renn (1981a), and a more specific review of the relevant psychological literature in Slovic, Fischhoff & Lichtenstein (1979).

The ideas presented in this chapter are drawn principally from four lines of research. The first attempts to describe how people perform probability judgments. The general thrust of this work is towards what Tversky & Kahneman (1981) call the psychology of choice. The second set of studies attempts to reveal the differences in such judgments among individuals drawn from different kinds of groups. This line of research is especially interesting because it sheds light on certain difficulties in communicating about technological hazards which derive from the differences in the way people perform relevant judgments. The third source of ideas in this chapter is research on the formation of beliefs and their relation to attitudes and behavior. Finally, a fourth line of research examines the possibility that many of the changes in attitudes and behavior towards technologies and their hazards may stem from changes in the basic values of large portions of the population.

Cognitive Attributes of Decision about Hazards

In thinking about response to a hazard, whether it stems from technology or some other source, it is important to recognize that uncertainty is involved in a fundamental and unavoidable way. Ravetz (1980) describes three cognitive aspects of thinking about risks: risks are universal, unavoidable, and conceptually uncontrollable.

A decision to act in response to a hazard, like every decision for action or inaction, interacts with all other areas of an individual's experience, for the individual tends to view each decision as yet another element in his total experience. Thus, the attempt to produce a taxonomy, a procedure for evaluation and a program of response to technological hazard is equivalent, in terms of individual reaction, to putting all human experience onto a scale for calculation and manipulation. The first cognitive attribute of risk is therefore that reaction·to hazards involves the full variety of life in its many dimensions of experience. An individual reacts only with reference to the totality of his experience, and this reaction in turn becomes a part of his experience.

The second cognitive attribute of risk is that they are generally incalcuable and often unimaginable. Consideration of major hazards, where damage resulting from occurrence would be very great but the likelihood of occurrence quite small, is often rather like taking a chance in a gigantic lottery. The outcome is viewed as depending upon luck rather than a calculated gamble. For more "ordinary" events, say the evaluation of the dangers involved in a particular automobile trip, the information available is generally inadequate for a proper calculation of the odds. General statistics about auto travel may be known, but usually not whether they apply to the route to be taken, the time of day, and the other particulars of this specific trip. For exotic, unfamiliar hazards, individuals seldom pretend to understand, or even imagine, the risks involved unless specifically asked to do so.

Finally, hazards are conceptually uncontrollable. One can never know whether enough has been done to prevent a hazard from occurring. Even after an event has occurred, it is impossible to be certain how much more would have had to be done to prevent it from occurring, or whether that level of activity would have been "reasonable".

One consequence of these cognitive attributes is that the decision about how to respond cannot be based simply on examination of cause-and-effect information about the chain of events from original source to ultimate effect, at least not when the latter includes human reaction. Whenever people's reactions play an important role, the analysis must necessarily be based on partial information. And as we have come to learn, such reactions are indeed important elements in the response to certain technological hazards.

One body of research relevant to this problem examins how people perceive, process, and evaluate the probabilities of uncertain events. In particular, the study of "heuristics" (simplifications imposed on information which enable inferences to be drawn from uncertain or missing data) is illuminating about how people make decisions in situations of incomplete or uncertain information. Tversky & Kahneman (1974) identified three heuristics people use in making probability judgments under uncertainty. The first is called "representativeness", and is applied in judgments of the following type: What is the probability that object A belongs to class B? What is the probability that event A originates from process B? What is the probability that proccess B will generate event A? In answering such questions people often rely on the representativeness heuristic, in which the probability is based on the degree to which A is representative of, or resembles B. When A is highly representative of B, the probability that A originated from B is judged to be high. On the other hand, if A is not similar to B, the probability is judged to be low.

The second heuristic typically applied in probability judgments is called "availability", and refers to the ease with which the event or class of events can be brought to mind. For example, one might judge the probability that a given business venture will fail by imagining the various difficulties it might encounter.

The final heuristic suggested by Tversky and Kahneman is called "adjustment and anchoring". It refers to the tendency to start from a preliminary value which is then adjusted to produce the final answer. The initial value, or anchor, may be suggested by the formulation of the problem, or it may be the result of partial computation.

Research suggests that the heuristic selected, the way it is applied, and the accuracy of the judgment it produces are all highly problem-specific. They may even vary with different presentations of the same problem. (Slovic, Fischhoff & Lichtenstein, 1977, p. 5-6) This variation in preference and assumption within the decision process has led to the formulation of a new model of choice, called prospect theory. (Kahneman &

Tversky, 1979)

Prospect theory provides an alternative basis for choice to that of expected uility theory, the most common model of individual choice in economics. The expected utility model presumes a set of axioms which provide criteria for rational choice. These include certain restrictions or definitional constraints including independent and exclusive options, and transitivity of preferences (these conditions have been formulated in a number of ways, see: Friedman & Savage, 1948; Luce & Raiffa, 1957; Markowitz, 1959). The choices by an individual conforming to the axioms can be described in terms of the utility of the various outcomes for that individual. The utility of an uncertain choice is equal to the expected utility of its outcome, obtained by weighting the utility of each possible outcome by its probability. When faced with a choice, the rational individual will prefer the prospect that offers the highest expected utility. (von Morgenstein, 1947; Raiffa, 1968; Fishburn,1970)

In prospect theory, outcomes are expressed as positive or negative deviations (gains or losses) from a neutral reference outcome, which is assigned a value of zero. Although subjective values vary among individuals, a common value function is S-shaped, concave above the reference point and convex below it. That is, the value function curves away from the 45 degree line (representing a one-to-one relationship in the change in gain/loss and derived value) towards the horizontal in the upper-right quadrant and downwards towords the vertical in the lower-left quadrant. This means, for example, that a gain from \$10 to \$20 is seen as more important than a gain from \$1010 to \$1020. Another property of the function is greater sensitivity to loss than to gain (i.e., the slope of the value function is greater below the neutral point than above it). This reflects the general tendency to be conservative in risk taking and the reluctance to accept a fair gamble.

The second major deviation of prospect theory from the expected utility model concerns the weighting of outcomes according to their likelihood. In expected utility theory the utility of an outcome is weighted according to its probability. In prospect theory, it is multiplied by a decision factor which is a monotonic function of probability but is itself not a probability. The essential characteristics of the weighting factor are that low probabilities are overweighted whereas moderate and high probabilities are underweighted, and that the effect of overweighting is greater than of underweighting. (Hypothetical examples of the value function and weighting factor are presented in Figure 5.1.)

Although the value function and weighting factor summarize a common pattern of choice, they are not universal. The preferencees of some individuals are not well described by an

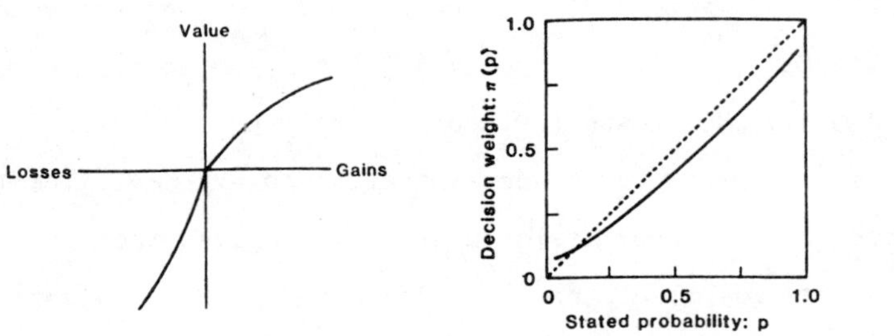

A hypothetical value function A hypothetical weighting function

Figure 5.1 Common Shapes for Value and Probability Weightings in Prospect Theory

Source: Tversky & Kahneman, 1981

S-shaped value function and a consistent set of decision weights. (Tversky & Kahneman, 1981, p. 454) Nevertheless, the pattern is common enough that the conclusions it suggests about preferences concerning alternative options should be considered when examining the cognitive aspects of reaction to hazards.

If both the vaule function and weighting factor were linear throughout, the preference order between options would be independent of the way options, outcomes or contingencies were formulated. But because of their nonlinearities, different formulations can lead to different preferences. For example, because the value function is steeper for losses than for gains, a difference between two options will appear greater if formulated as the disadvantage of one option rather than the advantage of the other. Similarly, probability insurance, which reduces the probability of loss by half, would be judged to be worth less than half the price of regular insurance that eliminates the chance of loss altogether. Such consequences of nonlinear value and weighting functions may explain the complaints of some analysts (presumably applying the assumptions of expected utility theory) that regulators (presumably consistent with prospect theory) irrationally expend too much money and effort to reduce hazards to extremely low levels. Prospect theory predicts greater value derived from reduction of loss at very low levels than does expected utility theory.

Judgmental Differences Among Groups

Evidence supportive of judgmental heuristics (and indirectly of prospect theory) has been collected not only using paper and pencil tests of hypothetical decisions (e.g., "Which of the following options do you prefer? A sure win of $30 or an 80% chance to win $45?" Tversky & Kahneman, 1981), but also psychometric surveys of views about various hazards. (Slovic, Fischhoff, & Lichtenstein, 1980) The latter research is of particular interest not only because it describes apparent differences among selected groups, but also because the selection of groups studied was designed, among other things, to reveal differences between "expert" and "lay" judgments. Group differences, and especially differences between lay persons and experts, appear of considerable importance in conflicts about technological hazards.

In one early study (summarized in: Slovic, Fischoff & Lichtenstein, 1980) three groups of lay persons (30 college students, 40 members of the League of Women Voters, and 25 business and professional members of a civic club, all from Eugene, Oregon) and one group of experts (15 persons professionally involved in risk assessment in the United States) were asked to rate 30 items. Included were activities (e.g., smoking, fire fighting), substances (e.g., food coloring) and technologies (e.g., railroads, aviation). Each was to be rated according to its present risk of death. The three groups of lay persons were fairly consistent in their ratings, but the experts differed markedly. (See Table 5.1.) The researchers interpreted the results to mean that the respondents believed that greater risk should be tolerated for more beneficial activities and that a double standard is appropriate for voluntary and involuntary activities. However, there also appeared to be a similar

Table 5.1 Fatility Estimates and Disaster Multipliers for 30 Activities and Technologies

Activity or Technology	Techincal Fatality Estimates	Geometric Mean Fatality Estimates Average Year		Geometric Mean Multiplier Disastrous Year	
		LOWV	Students	LOWV	Students
1. Smoking	150,000	6,900	2,400	1.9	2.0
2. Alcoholic beverages	100,000	12,000	2,600	1.9	1.4
3. Motor vehicles	50,000	28,000	10,500	1.6	1.8
4. Handguns	17,000	3,000	1,900	2.6	2.0
5. Electric power	14,000	660	500	1.9	2.4
6. Motorcycles	3,000	1,600	1,600	1.8	1.6
7. Swimming	3,000	930	370	1.6	1.7
8. Surgery	2,800	2,500	900	1.5	1.6
9. X rays	2,300	90	40	2.7	1.6
10. Railroads	1,950	190	210	3.2	1.6
11. General (private) aviation	1,300	550	650	2.8	2.0
12. Large construction	1,000	400	370	2.1	1.4
13. Bicycles	1,000	910	420	1.8	1.4
14. Hunting	800	380	410	1.8	1.7
15. Home appliances	200	200	240	1.6	1.3
16. Fire fighting	195	220	390	2.3	2.2
17. Police work	160	460	390	2.1	1.9
18. Contraceptives	150	180	120	2.1	1.4
19. Commercial aviation	130	280	650	3.0	1.8
20. Nuclear power	100[a]	20	27	107.1	87.6
21. Mountain climbing	30	50	70	1.9	1.4
22. Power mowers	24	40	33	1.6	1.3
23. High school & college football	23	39	40	1.9	1.4
24. Skiing	18	55	72	1.9	1.6
25. Vaccinations	10	65	52	2.1	1.6
26. Food coloring	—[b]	38	33	3.5	1.4
27. Food preservatives	—[b]	61	63	3.9	1.7
28. Pesticides	—[b]	140	84	9.3	2.4
29. Prescription antibiotics	—[b]	160	290	2.3	1.6
30. Spray cans	—[b]	56	38	3.7	2.4

[a] *Technical estimates for nuclear power were found to range between 16 and 600 annual fatalities. The geometric mean of these estimates was used here.*

[b] *Estimates were unavailable.*

Source: Slovic et al., 1980

application of double standards based on characteristics such as controllability, knowledge, faimilarity, and immediacy of the hazard.

A further study was conducted by the same researchers to clarify the results. 90 hazards, representing a very broad range of activities, were rated by a single group of lay persons (college students) according to 18 risk characteristics (the first study used nine characteristics). The results of this second study present a somewhat more complex picture of the factors influencing judgments of the risks involved with various hazards. They can be described by referring to the correlation matrix found in Figure 5.2. This figure shows that involuntariness is closely related to several other characteristics, including lack of control, global catastrophe, inequity, and catastrophe. For example, six of the most involuntary hazards (nuclear weapons, nerve gas, terrorism, warfare, nuclear power, and DDT) are also among the ten most catastrophic hazards. This suggests that much, if not all, of the observed aversion to involuntary risks may be due to other characteristics that are closely related to involuntariness.

The authors also suggest a plausible example to illustrate their doubts about the importance of involuntariness. Suppose you own and ride in two automobiles. One is for work, and when you ride in it you are chauffered. Thus, exposure to injury or death when riding in this car is involuntary and relatively uncontrollable. The second auto is driven by you for pleasure, and the risks you take in it are voluntary and somewhat under your control. Would you demand that technical safety standards be more stringent for the chauffered auto than for the one you yourself drive? Most people would not, illustrating (at least in

Table 5.2 Intercorrelations Among 18 Risk Chacteristics in the Extended Study

	Severity not controllable 1	Dread 2	Globally catastrophic 3	Little preventive control 4	Certain to be fatal 5	Risks & benefits inequitable 6	Catastrophic 7	Threatens future generations 8	Not easily reduced 9	Risks increasing 10	Involuntary 11	Affects me personally 12	Not observable 13	Unknown to those exposed 14	Effects immediate 15	New (unfamiliar) 16	Unknown to science 17	Many people exposed 18
1. Severity not controllable																		
2. Dread	.82																	
3. Globally catastrophic	.78	.83																
4. Little preventive control	.86	.72	.71															
5. Certain to be fatal	.80	.82	.73	.77														
6. Risks & benefits inequitable	.75	.76	.84	.63	.65													
7. Catastrophic	.77	.66	.77	.74	.67	.77												
8. Threatens future generations	.62	.76	.86	.52	.64	.81	.63											
9. Not easily reduced	.59	.67	.64	.67	.66	.59	.56	.59										
10. Risks increasing	.51	.63	.76	.48	.58	.76	.57	.75	.60									
11. Involuntary	.68	.58	.69	.52	.42	.77	.74	.62	.33	.44								
12. Affects me personally	.57	.64	.77	.43	.50	.71	.64	.78	.35	.67	.61							
13. Not observable	-.04	-.14	.04	-.19	-.28	.08	.00	.24	-.15	-.10	.33	.10						
14. Unknown to those exposed	.14	.05	.22	-.05	-.12	.28	.24	.35	-.20	.05	.63	.31	.79					
15. Effects immediate	.21	.15	.00	.34	.36	-.08	.11	-.29	.22	.01	-.25	-.14	-.87	-.77				
16. New (unfamiliar)	.32	.29	.32	.22	.20	.25	.18	.44	.13	.06	.36	.20	.63	.56	-.48			
17. Unknown to science	-.10	-.10	-.04	-.07	-.15	-.02	-.04	.02	-.23	.02	.13	.00	.43	.50	-.44	.32		
18. Many people exposed	-.04	.04	.23	-.14	-.11	.32	.23	.47	.03	.26	.34	.56	.37	.46	-.52	.07	-.01	

this case) the irrelevance of the involuntary component of exposure. The authors conclude that the apparent aversion to involuntary hazards may be mostly illusionary, caused by the fact that such hazards are often undesirable in more important ways, such as being inequitable or catastrophic.

The correlation matrix in Figure 5.2 reveals three sets of characteristics for the 90 activities rated in the study. The first 12 characteristics correlate more with each other than with the remaining characteristics. As a group, these characteristics may be said to reflect the degree to which the concequences of the hazard are particularly distasteful. The authors refer to this set of characteristics with the label "dread". A second set of characteristics (13-17) also correlate better with each other than with the rest. Collectively, they deal with the degree to which the sequence of cause and effect from the source of the hazard to the undesirable consequence(s) is understood. This set was labeled "familiarity". The last characteristic, the number of people exposed, stands alone. This characteristic was labeled "catastrophic potential" by the authors. In summary, this research of Slovic, Fischhoff and Lichtenstein suggests that lay judgments about the risks involved in particular hazards are based on these three factors: dread, familiarity and catastrophic potential.

In comparison, the judgments of experts in the above studies are much more closely related to the probability of death associated with each activity. (Fischhoff, Slovic & Lichtenstein, 1980) The authors also note that when lay persons are asked about fatalities, they give fatality estimates. But when asked about risks, they give something else.

Other research has also indicated differences among

response groups. For example, Otway & Fishbein (1977) examined differences of persons favoring and opposing nuclear power. They found that the pro-nuclear energy group based their opinions on such factors as the "economic and technical benefits" and "environmental/physical risks" whereas the anti-group considered "psychological aspects" and "sociopolitical implications" to be of greater importance. A later study found that government officials differed from the public in terms of their assessment of the psychological aspects. (Thomas, Swaton, Fishbein & Otway, 1980)

Vleck and Stallen (1981) interviewed individuals in three age groups living at systematically varying distances from an area of heavy industrialization. Multidimensional scaling analysis resulted in three pairs of cognitive dimensions underlying individual judgments of "riskiness", "beneficiality" and "acceptability". The interrelations among these cognitive dimensions appeared to be consistent over several aspects of such judgments. Differences appeared in secondary dimensions of riskiness, beneficiality, and acceptability, however, which could to some extent be traced to systematic differences in group characteristics. No relationship was observed between judgments of overall riskiness and beneficiality. Scale values of beneficiality did, however, strongly correlate with dimension values of acceptability, which suggests that riskiness plays a secondary role is the judgment of the acceptaability of risky activities.

Renn (1981a; 1981b) compared groups with respect to several criteria. His results indicate that those favoring nuclear power have less homogeneous and consistent attitudes (e.g., while exhibiting a strong emphasis on economic benefit, they are aware of the hazards involved) than those in opposition. Opponents appear to have much less confidence in experts than do supporters of nuclear power, but both groups consider scientists from universities and research institutions to be more reliable that experts from political organizations, industry, churches or unions. Finally, people have more positive attitudes towards energy technologies with which they have some experience; for example, when they live in the vacinity of such a technology. Renn's major conclusion is that judgments about energy sources do depend upon the inherent hazards and their associated risks, but that specific beliefs about the advantages and disadvantages of the technology involved have a strong effect.

Eisenhardt & Krebsbach (1978) examined the opinions of citizens, experts and politicians about themselves and each other with respect to planning processes for large technological facilities. In small (6-8 person) directed group discussions, separate groups of citizens, experts and politicians were asked to portray themselves and each of the other groups in terms of their roles in the planning process. All groups saw citizen participation, although formally a part of the process, as little more than a farce. In general, it was felt that decisions had already been made before the citizens become involved, and that they usually become active only in order to hinder something. Both politicians and citizens see experts as difficult to understand, although to different degrees. None of the groups was willing to accept responsibility for the overall decision. Citizens felt that in the end the decision should be made by experts. Experts saw the decision as the responsibility of politicians, who in turn felt that they are always dependent upon experts when it comes to technical matters. Each group had different views of each other and their respective roles in the planning process.

Beliefs, Attitudes and Behavior Towards Technology

Up to this point in this chapter, human reaction to technological hazards has been described in terms of a model of the way people interpret uncertain and partial information in making choices. Another model which has been used in the study of reaction to technological hazards is that of belief, attitude, intention and behavior developed by Fishbein & Ajzen (1975). According to this model, an attitude may be defined as an evaluative judgment of the degree one likes or dislikes some person, object, concept or symbol. Attitudes are built upon beliefs held about the subject, which are learned relationships between the object and its attributes or characteristics. For example, a Porsche sportscar (the object) is fast (an attribute). The strength of a belief is related to the intensity of the learning process, which is an important influence on the relationships among beliefs, attitudes, intentions and behavior. (Otway, 1980)

As we develop, we learn through experiencing the relationshps among the objects and concepts that form our environment. In this way we gradually build up a view of the world and the way it works. This view will be rather precise, insofar as the perceptual inputs are based on personal experience. It follows that the experiences various people have will be different, in part due to the restrictions and variations imposed by the physical and social environments in which they find themselves. Since one rarely questions the validity of one's own senses, these descriptive beliefs, based on personal sensory experience, will be held with maximum certainty. We also

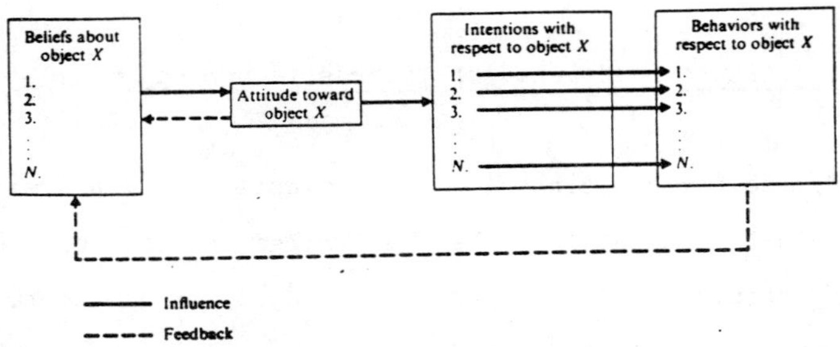

Figure 5.3 Schematic Presentation of a Conceptual Framework relating Beliefs, Attitudes, Intentions, and Behaviors with respect to a Given Object
Source: Fishbein & Ajzen, 1975, p. 15

receive information from others which, to the extent that it is accepted, enters into the information system of beliefs. These beliefs are further extended by inferential processes based upon the beliefs already held. A belief system thus acquired allows one to go beyond the information provided in a given situation to make predictions (inferences) about events and relationships that have not yet been experienced. The relationships among the various elements are shown schematically in Figure 5.3.

Typically a relatively large number of beliefs underly any given attitude. Even when the individual has a single, clearly formulated attitude about a certain object, that attitude predisposes the person to behave in a way consistent with that attitude only in terms of the totality of his behavior with respect to that object. Taken together, these behaviors will be consistent with his attitude. Thus, attitudes about technologies and their hazards may not be exact predictors of behavior in a given instance, but should indicate general predisposition which determine behavior over time.

This model suggests that different kinds of information will affect individuals differently. For example, a device on a factory smoke stack that removes visible exhaust might change rather quickly the beliefs of the local residents that the factory pollutes the air, because the change is observable by their own sensory perception. But safety-related technical information is highly abstract and not subject to personal confirmation. Provision of such information can even stimulate the formation of inferential beliefs which are exactly to the contrary if the source of information is not accepted or is mistrusted (e.g., it must be a dangerous process if the government says that the level of danger is acceptable).

Rather dramatic support for the importance of personal sensory perception in comparison to abstract information comes from an unusual source: experiments with students' reactions to demonstrations of "psychic" abilities. (Hofstadter, 1982) Students first filled out a questionnaire to obtain an estimation of their prior beliefs about psychic powers. Then a "frighteningly impressive" demonstration of psychic ability was performed. Following this, half of the students were "dehoaxed". This process, lasting only two or three minutes, consisted of showing how two of the three tricks worked, together with a confession that the remaining one of the baffling stunts was also a trick. This was followed by a 50 minute lecture in which the secrets of professional mindreaders, commonsense estimates of apparent coincidences ("oddmatches"), and other arguments to undermine psychic phenomena in general were revealed. The other half of the students heard only the lecture. The students were then polled about their belief in various kinds of psychic phenomena. The classes that had seen the dehoaxing demonstration had a far lower belief in such phenomena than those that had simply heard the lecture. A three minute revelation that they had been fooled was more powerful than an hour-long lecture in producing skepticism.

Another experiment with reaction to psychic phenomena shows how tenatiously people hold on to the beliefs they have obtained through their own sensory perception. (Hofstadter, 1982) In two classes a magician was introduced as a graduate student "interested in the psychology of paranormal or psychic abilities (who has) been working on a presentation of his pyschic abilities." The instructor also explicitly stated that he was not convinced of anyone's psychic ability. In two other classes, the performer was introduced as a graduate student "interested in the psychology of magic (who has) been working on developing a presentation of his magic act." Only a few students seemed to believe the instructor's description of the performer as a magician in the two classes where he was introduced as such. Two more classes were given the added warning "(the performer) will pretend to read minds and demonstrate psychic abilities, but ... does not really have psychic abilities, and what you will be seeing are really only tricks." In spite of all the forewarning, more than half of the students expressed the belief afterwards that the performer did in fact possess psychic powers. The next question was whether magicians could do exactly what had been demonstrated. Virtually all students agreed that magicians could, but only a few reversed their estimate of the psychic powers of the performer they had seen. 55% still believed he had psychic powers. They were then asked how many people claiming psychic powers were actually fakes. Consensus was that three out of four "pyschics" were in fact frauds. But even after supplying such negative information themselves, only a few wished to revise their estimate of the performer's psychic abilities, reducing the proportion of those believing in his powers to 52%. The strength of beliefs based on one's own sensory perception can withstand even abstract but contrary information which the individual himself holds true in general.

Applications of the Fishbein and Ajzen model to technical issues reveal not only the differential affects of various kinds of information, but also the complexity of attitudes involved. For example, one study contrasting those most in favor of the use of nuclear power with those most opposed to its use showed that people treat the risks and benefits involved independently rather than as lying along a single axis, as might be inferred from general notions of cost-benefit analysis. (Otway, 1980) The study further showed that people differentiate among various kinds of hazards. It showed the largest contributions to the attitudes of those favoring nuclear power to be made up of factors associated with the technical and economic benefits, and with the environmental and physical consequences. In contrast, psychological aspects and sociopolitical implications of the use of nuclear power dominated the attitudes of those opposing its use. This suggests that the two groups, while often appearing to debate the risks of nuclear power, may well be referring to entirely different things.

Otway (1980, p. 43) observed the following points based on this research.

> 1. Because technical information can seldom be verified by one's own senses, it would not necessarily be expected to play a dominant role in the formation or change of public attitudes on technical issues. Therefore, technical safety studies undertaken with the idea of providing "hard facts" to influence public

opinion are unlikely to achieve the desired result.

2. Perceptions of risk are multiply determined and would not be expected to agree with statistical estimates, or even measurements, of risk level.

3. Because opposition to technologies is often due to factors other than risk in its conventional sense, numerical guidelines for "acceptable risk" based upon statistical data on the experience of existing risks (a normative approach) would not be expected to gain public acceptance.

4. Risk estimates and historical risk statistics _can_ be used in the regulatory process, but only to help determine an _upper limit_ of risk that must not be exceeded. _It is a necessary, but not sufficient_ condition that the estimated risk of new technologies be less than this upper limit.

5. Points 2 and 3 imply that the definition of acceptable risk will vary from case to case. The _process_ by which criteria are derived is likely to be more important to their social acceptability than are any numerical guidelines that might result from the process.

These points are consistent with the position taken in this study. The principal focus of this study is on the processes by which judgments about the consequences of technological hazards and appropriate responses to them can be organized. The five points listed above provide a brief summary of the role of information and the importance of process in issues where public opinion plays an important role.

Values and Value Change

Certain technological issues, the most notable being the debate about nuclear power, clearly demostrate that attitudes about the technologies involved and their hazards cannot entirely explain the conflicts involved. In general, there appears to be little correspondence between the degree of hazard involved and the level of acceptance. Evaluation and acceptance of technological hazards depends, among other things on interests, goals, and values--all controversial, and all which may change over time. (Bechmann, Frederichs & Gloede, 1981)

A great deal of social science research on the application of new technologies has focussed on questions of hazards and their associated risks and the related problems of acceptance. Much of this research has employed the methods of social psychology, for example as described in the previous sections. However, other research on elections and political allignment has demonstrated both that value orientations are important influences on certain attitudes and actions (Pappi & Laumann, 1974), and that a major shift in value orientation may be occurring in the population as a whole (Inglehart, 1971). Thus, part of the conflict may arise, at least for some people, not from the hazard associated with the technology but rather from a value orientation which opposes altogether the use of that kind of technology. Examination of only the attitudes held about a technology and its associated hazards may underemphasize the importance of changes in value orientation, or may not reveal them at all.

About a decade ago, Inglehart (1971) hypothesized that the

basic value priorities of Western publics had been shifting from a "materialist" emphasis towards a "post-materialist" orientation--from giving top priority to physical sustenance and safety towards heavier emphasis on belonging, self-expression, and the quality of life. (In other words, a shift towards "self-actualization" as described by Maslow, 1954.) The theoretical basis for this supposition is based on two key hypotheses: 1) a scarcity principle, in which one places the greatest subjective value on those things that are in short supply; and 2) a socialization hypothesis, in which one's basic values reflect the conditions that prevailed during one's preadult years. (Inglehart, 1981) The recent economic history of advanced industrial societies has significant implications in the light of the scarcity hypothesis. For they are a remarkable exception to the prevailing historical pattern: the bulk of their populations have not lived under conditions of hunger and economic insecurity. The scarcity principle suggests that such a period of prolonged prosperity should lead to an increase in post-materialistic values. However, the socialization hypothesis insures that there is no simple one-to-one relationship between economic conditions and prevalent value orientation per se. Fundamental value change takes place gradually. In the general population, it generally occurs as a younger generation replaces an older one in the adult population. But a considerable time-lag would be involved. After a period of sharply rising economic and physical security, one would expect to find substantial differences between the value priorities of young adults and their elders. Ten or fifteen years after such prosperity begins, the age cohort that had experienced prosperity during their formative years would enter the electorate. Perhaps another decade would pass before that group would begin to occupy positions of power and influence in their society, and that long again before they reached the level of top decision makers.

Recent surveys suggest that a majority of those of 35 years of age and less who occupy jobs that lead to top management and top civil service posts express post-materialist values. By the end of the 1970s, the post-materialist value orientation had become a force to be reckoned with among technocrats and professionals in their thirties and forties. This does not necessarily mean that it will dominate political activities in Western societies. Those with the post-materialist orientation remain a numerical minority in the overall polulation, better equipped to attain their goals through bureaucratic institutions or the courts than through the electoral process. (Inglehart, 1981)

The importance of the emergence of the post-materialist value orientation among such key parts of the population is clearly revealed by examining the debate about environmentalism and economic growth. Until relatively recently, it was taken as self-evident that economic growth was inherently good; although there was considerable disagreement about the allocation of benefits, economic growth was encouraged by both management and labor, capitalist and communist. Only recently has the environmental movement questioned this assumption, claiming that economic growth does not always justify its human and environmental consequences. When environmentalism raises questions of environmental quality versus economic growth, it pits post-materialist priorities squarly against materialistic ones. (Inglehart, 1981) It is significant that such pressures have nct disappeared during the period of economic strain following the energy crisis in 1973. Environmentalists have retained their views, perhaps become even more strident, despite the energy crisis, general economic malaise, and emergence of a vocal opposition to environmentalist goals. This supports the contention that environmentalism is based on a fundamental value orientation that will change very slowly if it changes at all.

Like the confrontation between environmentalism and economic growth, the debate about nuclear power reflects a clash of world views. Materialists see the use of nuclear power as desirable to the extent that it contributes to economic growth and full employment. For them, highly developed science and industry symbolize progress, prosperity and security. Among post-materialists, however, nuclear power tends to be rejected not only because of its potential dangers but also because it represents big business, big government and big science--bureaucratic organizations that are disliked because they are inherently impersonal and hierarchical, minimizing individual self-expression and human contact. (Inglehart, 1981)

However, a note of caution must be included concerning this postulation about emerging value changes. While the overall emergence of post-materialist values appears plausible on the societal level, certain inconsistencies appear when one compares different groups within a single society. For example, the scarcity principle states that people will value more highly those things in short supply, and this is then used to explain a shift in focus from economic security to post-materialist goals. This appears to correspond to general developments. But the value change formulation, in terms of the scarcity principle and the socialization hypothesis, fails to explain why the individuals who appear to exhibit the greatest shift to post-materialist values tend to come from well to do, upper middle class families. Exactly these families have experienced relatively cleaner environments and better opportunities to pursue self-actualization as compared with their poorer counterparts. A combination of the Inglehart formulation of shift in values on the basis of these two hypotheses and the postulation by Maslow that people first satisfy simpler, more elemental needs and then shift to more abstract and obtuse rewards might suffice to explain differential value shifts among different social groups. But it must be concluded that at this time the literature on value change leaves a number of questions unanswered. The postulated shift in societal values cannot be rejected out of hand, but the evidence appears mixed.

Summary and Conclusions

This chapter addresses individual reaction to technologies and their hazards. It briefly reviews several different lines of research which shed light on the way people might react.

Because of the complexities of technological hazards, people can be expected to use one or more psychological heuristics to simplify the decision about how to respond. It also appears that people commonly attach greater importance to potential losses than to potential gains, and give disproportionate attention to possibilities with very small liklihood. Thus, the intuitive evaluation represented by an individual's reaction to technological hazards may incorporate conclusions somewhat at variance with those that would be derived using the traditional assumptions of expected utility theory.

A second general conclusion is that evidence which an individual can observe or confirm with his own senses will likely be more convincing than more abstract information, no matter how well documented. People, not irrationally, tend to judge new information in terms of what they already know. The things they know best, the beliefs they usually hold most firmly and change only grudgingly, are those based on their own sensory experience.

Finally, it may be that the basic values of a substantial portion of the population give less importance to material things, and consequently to traditional economic progress. This would lead to major differences of interpretation, not only in the determination of whether particular attributes of a technology should be viewed as positive or negative, but also in the degree of importance given to various factors.

6 RESPONSE TO TECHNOLOGICAL ISSUES

Protest against technologies and their use is by no means a new phenomenon. Many of the technologies developed during the industrial revolution and afterwards have met with varying degrees of opposition. Recent years have seen opposition to rural electrification, water fluoridation, supersonic air transport, contraceptive devices, and nuclear power, to name but a few examples. (Lawless, 1977) Some such protests have been quite successful--further development of nuclear power appears stymied in many countries, use of DDT has been severly curtailed, the SST (supersonic transport) appears to be an oddity rather than the harbinger of the future in commercial aviation.

But there have been even more fundamental consequences of opposition to unfettered development of technology. In most industrialized countries government regulations have been established, usually within the last ten to fifteen years (although some are far older), which are designed to serve as checks on unlimited technological development. Such laws and regulations typically require the examination of environmental impacts of proposed activities, of the safety of products and processes, or of the interests of workers or consumers. But response to technological hazards has not been limited to those who "oppose" technological development. Business enterprises have also recognized the need for joint action. In the area of chemicals, for example, toxicology research programs have been established by industry in several countries and western Europe. Such recognition of the advantages of cooperative efforts on the part of industry is also a part of response as treated in this chapter.

Response to technological issues--that is, the action which is undertaken as opposed to the beliefs and attitudes which people hold--must be described in terms of collective action as well as the behavior of individuals. This is especially so because the scale of modern industrial activity often means that individual response to the problems it creates can be effective only in quite rare circumstances. Of course, examples can be found of highly motivated individuals taking on the corporate giants, and sometimes even achieving a degree of success. (Knoedgen, 1981) But it is much more common to find motivated people working through an organized group. Such groups can be more or less formally organized, and range from general purpose organizations like The Friends of the Earth or the World Wildlife Fund to groups that focus on specific goals such as the construction of a particular facility. Citizen opposition to the additional runway at Frankfurt airport or to the proposed third London airport are good examples of the latter. Sometimes groups are organized around a charasmatic personality, such as Ralph Nader, but often they simply result from the similar views and goals of their members.

This chapter presents a set of categories that can be used to describe collective response to technological issues. These categories rely heavily (but not exclusively) on concepts developed to describe actions of organizations at the international level. There are two principal advantages to this. First, the multiplicity of soverign polities at the international level requires reference to complicated decision processes, so that the concepts used are capable of dealing with complex situations. Second, the essentially arbitrary nature of many so-called "technical" standards are clearly revealed in international arenas. Although it is usually claimed that technical standards are based on objective scientific and technical analysis, comparison of standards in different countries suggests that they are a result of trade-offs among many technical, political and administrative factors. It is difficult to believe that Frenchmen differ from Italians in sensitivity to air pollution to the extent that the respective technical standards in the two countries would suggest.

The ideas presented here do not rely extensively on the standard economic treatment of group behavior, usually referred to as the economic theory of clubs. A club has been defined as "a voluntary group deriving mutual benefit from sharing one or more of the following: production costs, the members' characteristics, or a good characterized by excludable benefits." (Sandler & Tschirthart, 1980, p.1482) The economic analysis of clubs is relatively recent, generally described as derived from seminal works by Olson (1965) and Buchanan (1965).

In The Logic of Collective Action, Olson (1965) describes group action as the provision of a collective or public good (i.e., something which, once provided, is equally available to all whether or not they contribute to its provision, and whose consumption by one individual does not limit its consumption by any other). As long as some individual receives more benefit than it costs to produce a public good, there is clear incentive for its provision. Olson calls such groups "priviledged groups". But for very large groups where the benefit to any one individual is the equivalent of only a small portion of the cost of providing the good, there is no incentive to provide it. A public good would be provided in such a "latent group" only when its provision is tied to some separate, selective incentive (something which is not indiscriminate like a public good, but which can be withheld when the individual does not contribute). Smaller groups, where it is possible to observe whether or not each individual contributes his bit, but where no individual receives sufficient benefit to produce the public good himself, are called "intermediate groups". If a public good is provided in an intermediate group, the analysis shows that it will be provided at a sub-optimal level (each individual receives less of the good than he desires). An optimal level of public good would be provided only when the marginal cost of producing it (the cost of the last unit produced) is distributed among the individuals exactly as is the marginal benefit (the additional benefit derived from the last unit).

A common phenomenon associated with public goods is called the "free rider" problem. There is an incentive for the individual to benefit from the public good, since it is available to all, but to avoid contributing to its production. Under the conditions of an intermediate group, where no one individual receives sufficient benefit to bear the cost of providing the public good, none of the good would be provided by economically rational actors. However, although examples of freeridership abound (e.g., hitchhiking, photocopying articles rather than purchasing journals, listening to television without paying license fees, evasion of taxes), there are also examples in which the opposite seems to occur. People vote, even though the outcome is relatively independent of their own behavior. Queues form (at least in some societies) even though the individual could better his position by crowding. And there is a whole set of activities in which individuals may be said to produce an excess of some public good: Ghandi and his followers, the hunger strikers in Norther Ireland, the Red Guard in Italy, the activitists in the "Solidarity" union in Poland. Despite the dissimilarity of their causes, in each of these cases a number of persons experience a degree of hardship from which they can hardly expect to benefit to a degree which justifies the costs they incur.

The standard approach to correcting for free rider activity, or for dealing with activity in general which involves external affects on others than the actor producing them, is to provide an organization charged with allocating the cost of providing the public good among those who experience its benefits. The greater part of the economic theory of clubs deals with the determination of such cost allocations, and with the optimal size of sharing groups. (A review and summary of the economic theory of clubs may be found in: Sandler & Tschirhart, 1980.)

Buchanan (1965) provided the basic theoretical underpinnings for the analysis of voluntary clubs, which he characterized as follows: homogneous members equally sharing both the public good and the costs associated with it, costless exclusion of nonmembers, and voluntary, decentralized (all members are identical and therefore need no central direction) operation without transaction costs (the costs associated with communication, negotiation, etc. rather than production of the good per se). New members should be added to a group so long as the marginal benefits (usually decreasing due to crowding) exceed the marginal costs (usually decreasing due to cost sharing).

The Buchanan analysis assumes a perspective that may be

said to focus within the group. That is, it considers only the benefits and costs accruing to the members of the group. Subsequent analysts have extended this perspective in a number of ways. Perhaps the most important is to incorporate both members and nonmembers in the analysis. But the analysis has also been extended to include heterogeneous members (differential utilization of the good and the associated allocation of costs through "tolls"), discrimination against some group members, transaction costs among members, and exclusion costs (costs associated with the erection, maintenance, or existence of a mechanism to limit club utilization or membership). The utility of a public good to a group member depends positively on the consumption rate of the good and negatively on congestion and crowding.

In an intriguing extension of the analysis of clubs, Coleman (1982) refers to the fact that with respect to public goods, a number of people's interests are likely to be satisfied by the same outcome. Therefore each might have an incentive to reward some individual to produce the good, in addition to his willingness to contribute something directly to the provision of the good. The economic theory of clubs argues that an individual will contribute to the provision of a public good so long as the marginal utility resulting from his contribution is greater to him than his marginal cost. In the case of an extreme public good, his marginal utility is near zero, so that his willingness to contribute is low. However, if the cost of providing a reward to another individual is sufficiently low, then it is possible not only to provide the good, but to exceed the amount that would be provided if the marginal cost were distributed exactly as the marginal benefit. Coleman (1982, p. 11) describes the following different situations concerning the combination of direct benefits and rewards from third parties.

1. The marginal cost of action is greater at all levels of effort than the direct marginal benefit from his own efforts plus the marginal benefit through rewards from others. There will be a total absence of activity and no provision of the public good. All will behave as free riders.

2. The marginal benefit each experiences directly from his action, together with the marginal benefits each experiences from rewards from the others, contingent upon his action, are greater than marginal costs at zero level of activity, but are not sufficiently great that the public good is provided at optimal level of activity. Here there is some free rider behavior, but only enough to depress the level of activity.

3. The combined marginal benefits from his own action and from the rewards provided by others is sufficiently great that <u>more</u> of the public good is provided than would be the case if costs were allocated so that the marginal costs equalled the marginal benefits for all actors. This is the case of the zealot.

Coleman goes on to point out that one of the more interesting aspects of such a system of mutual rewards for activity directed towards achievement of a collective goal is that each actor experiences a larger total amount of reward than when the direct benefits are restructured so as to fall to those who bear the costs of provision. "Many accounts of communes and of religious communities describe the extreme sense of inner peace and good feeling." (Coleman, 1982, p. 12) Interactions among the various actors can provide incentives that are just as strong, or perhaps even stronger, than the direct benefits from the consumption of the public good.

The analysis provided by the economic theory of clubs is not inconsistent with the position taken in this chapter, but it focusses on different issues. The principal topics in this chapter are neither the optimal size of the group nor the optimal level of cooperation. Rather, attention will be given to more primitive or fundamental questions: What is it that motivates cooperative and collective action in the first place? How might one stimulate the formation a collective group? Which approaches might be best for coordinating activities within such a group? In Chapter 4, group action was described as an aggregation of individual activity, and much of the discussion was about how the individual can make use of groups to pursue his or her own aims. Group action was not treated as a public good in the sense used in the economic theory of clubs. To use the termonology of the latter, the interest here is in the identification of the selective incentives that might be used to encourage contribution to public goods (a good example is the provision of attractive magazines to members of environmental groups like the Sierra Club or the World Wildlife Fund, resulting in very large membership which is then used as a base for lobbying and other poltical activities), in the determination of the kinds of common interests which can provide the basis for a system of mutual rewards such as that described above, and in the ways activities can be organized and coordinated so as to overcome the transaction costs associated with large, diverse collective groups.

As was mentioned earlier, these concepts may be applied to actions on the part of those who "produce" technological hazards as well as those individuals and groups interested exclusively in the reduction or elimination of such hazards. The concepts put forward in this chapter are well suited to this purpose. Those derived from the analysis of "international regimes" are commonly used in this manner. Indeed, large enterprises (usually multinational corporations) are often important participants in international regimes. For example, the major oil companies have made substantial contributions to the current measures for dealing with oil pollution at sea. To be sure, they are well aware of their respective economic interests, and the structure of petrochemical industry and the shipping industry are clear influences on their contributions. Yet they also act cooperatively (Sibthorp, 1975; M'Gonigle & Zacher, 1979), and the concepts presented in this chapter can describe their behavior and those of international organizations and governments with respect to oil pollution at sea.

Furthermore, the approach to strategy presented in this chapter was actually developed to describe business enterprises. The concepts in that approach clearly refer to elements based on the way an enterprise interacts with other enterprises. But it does so not only in terms of competition, but also of cooperation. One could say that the structure of the industry provides the starting point, but that certain strategies required by current conditions require the restructuring or creation of new interactions, including those among enterprises as well as with other kinds of organizations.

The presentation in this chapter begins with an examination of the parameters of choice, describing both the boundaries within which such activities take place and the principal elements which interact. Next, the basis for common response, interdependence, is addressed. This sets out the most common reasons for collective action. After this, an analysis is presented describing the general strategies avaialable for an actor who must consider the interest and behavior of others in his choices. Finally, instruments for coordination of separate or collective activites are described.

Thus, the concepts presented in this chapter enable a description of response to technological hazards which emphasizes collective or joint action. Standard descriptions of management or coordination are eschewed in favor of an approach which relies on interdependence among different entities to provide the basis for collective and cooperative action. This commonality of problem or opportunity is viewed as the principal legitimation for intervention, and provides a clear connection to the conceptual framework in Chapter 4.

The Parameters of Choice

Technical parameters, revealed through natural science and engineering, are the primary determinants of the nature of a technological hazard and its consequences. But when one is interested in managing a situation or process rather than understanding the nature of the hazard, then political constraints become equally important, if not predominant. For example, the response that maximizes long run fish catch may reduce employment in the short term, or pollution standards which improve environmental quality may reduce trade potential. Choosing among alternative responses to technological options is not only a technical question, it is also a social choice.

The introduction of social and political elements into the choice of response to technological issues adds complexity to the decision calculus. This chapter is devoted to that extension, and this section begins the process by drawing the boundaries for the choice and attempting to reveal the additional factors that must be considered.

Response to technological issues cannot be treated as a simple matter of stimulus and response. Rather it should be seen as the product of the intersection of three different sets of competing influences. (Ruggie, 1975, describes international response to technology in terms of the first two axes presented here.)

Along the first axis is found the tension between science, heavily informed by consensual knowledge of cause and effect relations, and politics, heavily informed by normative purpose, negotiated priorities and available capabilities. Innumerable case studies of technological conflicts (e.g., Nelkin, 1978; Nowotny, 1979a, 1979b; Green & Hudspith, 1979) or of the decision making structures about technological development (e.g., Kueppers & Nowotny, 1979; Nelkin & Pollak, 1980) as well as of international organizations with responsibility in technological matters (e.g., Skolnikoff, 1972; Cox & Jacobson, 1973) document this interaction between scientific knowledge and political aims. The give-and-take between scientific and political influence is perhaps the most fundamental characteristic of the choice of response to technological issues.

Along the second axis is the tension between the need to respond collectively to problems and opportunities on the one hand and the desire to maintain autonomy and flexibility of action on the other. Organizations representing collective approaches to technological issues can be found every level, and in both the public and private sectors. At the international level, for example, the OECD (Organization for Economic Cooperation and Development), various United Nations organizations (especially the World Health Organization, the Food and Agriculture Organization, and the United Nations Environmental Program), and the European Community all have efforts which constitute collective responses to technological issues. Nongovernmental efforts are also to be found in the International Labor Office, in a variety of industrial associations such as the European Chemical Industry Ecology and Toxicology Center, and in special interest groups like Friends of the Earth. But in addition to the obviously large motivation to undertake cooperative and collective action, there is also a great deal of interest in maintaining autonomy. This interest can sometimes be said to be "pro-active". That is, it is often legitimate to claim that the most efficient control, say, of a particular pollutant takes account of the particularities of each separate source. It might be most efficient to apply different control technologies in different situations. But in practice, it is difficult to distinguish such an argument from one which is essentially "defensive" in character, aimed only at reducing the degree of external control. Autonomous response can also mean considerable freedom from having to do anything at all. Reasons for undertaking collective action are described in the next section. But these reasons are generally balanced against the desire to maintain autonomy and freedom of action. This tension between collective and autonomous action is a second characteristic of response to technological issues.

The third axis corresponds to the tension between the need to stimulate the development and use of new technologies and innovations on the one side and the necessity of coping with the consequences of using them on the other. The debate about pre-market testing of drugs in a classic example. Everyone wants to avoid a repeat of the thalidomide tragedy, in which an unanticipated side-effect of the sleeping pill in pregnant women led to deformed babies. The need for testing prior to marketing is not contested. However, it is argued that the extent of current testing requirements means that safe drugs reach the market far too slowly and at too great a cost. (Schnee, 1979) The tension between the stimulation and the control of technological development is clearly a central issue in the current discussion of "deregulation", and constitutes the third fundamental characteristic of the choice of response to technological issues.

The response to technology may thus be seen in terms of the tensions along these three axes. Taken together, they may be said to establish the parameters for the choice of reponse. Each response to a technological issues will represent some combination of the interactions along, and among, these three axes.

Interdependence

Of critical importance in the examination of collective, or organized, response to technological issues is the nature and degree of interdependence among individuals and groups in terms of the issue in question. This is because the basic justification for collective action is that the interdependence involved creates the possibility of achieving otherwise unobtainable advantages through cooperation. There are several types of interdependence which can serve as the basis for collective action, and which are therefore relevant for describing response to technological issues: cognitive, opportunity cost, constraint, and deprivation. (Ruggie, 1975, pp. 561-564)

Cognitive interdependence describes a situation in which it is recognized that a collective situation exists for which isolated action would likely be mutually inefficient, whereas collective action would likely be mutually beneficial. Cognitive interdependence may be said to the the principal underpinning, for example, of the European Community. The Treaty of Rome, as well as the common agricultural policy, (which help provide the base for the Community) recognize that coordination in industries like iron and steel or agriculture would benefit the participating countries and enable certain developments that would probably be impossible without such cooperation. Sometimes it is felt that mutual analysis of the issues involved will contribute towards appropriate response, as may be illustrated by many OECD efforts which simply produce recommendations and do not attempt to impose their solutions on the participating countries.

Cognitive interdependence may also be said to be one of the principal motivations behind the recognition of the "power in numbers" of membership in groups like the Friends of the Earth. An individual, or smaller group, is normally less politically effective than a large group. In banding together and undertaking collective action, more can be achieved, or it can be achieved more easily, than through individual action.

The second type of interdependency, opportuity cost, can constitute a common motivation among individuals or groups as follows. The pursuit of one objective may require other objectives to be reduced or foregone when undertaken individually, whereas collective action might alleviate the burden of a particular objective so as to allow resources to be allocated elsewhere. Recogntion of opportunity cost interdependence has provided the basis for cooperation in Europe on projects concerning high technology. Examples include the Concorde SST, the Airbus, the Toronado fighter plane, the European Space Agency, and CERN (a joint high energy physics laboratory). Opportunity cost interdependence is not limited to areas under direct government control, as these examples might suggest. It is also the basis, for example, of joint ventures in the automobile industry, where Daimler-Benz and Volkswagen share research costs, and Peugeot, Renault and Volvo jointly developed and manufacture an engine. Opportunity cost interdependence could be used to describe the phenomenon of "production sharing", where different steps in a manufacturing process are performed in different countries, depending on the avaialability of labor, capital, and other factors. A classic example is the tanning of American cowhide in Brazil, working it up into uppers in the British Virgin Islands and into soles in Haiti, and subsequently into finished shoes in Barbados or Jamaica for sale in Europe or Puerto Rico for sale in America. (Drucker, 1980)

Contingency interdependence refers to the recognition that action by one individual or oganization raises uncertainties for others in terms of the ability to pursue their own affairs, whereas collective action might provide an early warning and enable the possibility of precluding undesirable conflicts. In economic terms, contingency interdependence may be said to underly the formation of a cartel. Preconditions include that the number of units involved be small enough that one can keep track of them and that the consequences of their actions be traceable. Another response to contingency interdependence is the formation of professional associations or other informal meeting grounds for those involved. Professional organizations, chambers of commerce, industrial associations, and so on owe their existence in part to their ability to provide an informal setting in which various parties can meet and discuss emerging problems before having to address them formally within their respective organizations. Contingency interdependence also sometimes provides the motivation for bi-lateral treaties between governments.

Constraint interdependence refers to the situation in which individual action precludes the possibility of others engaging in that activity, whereas joint action might provide mutual benefits. The most common examples of constraint interdependence involve the procurement of natural resources. For example, constraint interdependence in the fisheries industry has stimulated various international management schemes aimed at controlling the catch of the overall industry, protecting all participants from unfair practices by one or more groups.

Another example is the control of the airwaves. Use of a given frequency by one broadcaster precludes its use by another (at least within a certain distance). The effective control of broadcasting requires cooperative action on the part of the relevant organizations, including both governmental agencies and public or private broadcasters in many countries. But in this example, the constraint interdependence is quite strong. Without cooperation, chaos would result.

Deprivation interdependence refers to the situation in which autonomous action deprives others of goods that rightfully belong to them, whereas joint action could allocate such goods for mutual benefit. One example is fishing in the territorial waters of another country (which is different from the constraint interdependence, which refers to the impact on the overall catch). A second example is acid rain. The increased acidity of rivers and lakes in Scandanavia is attributed to emissions from factories in Great Britian and Europe. Although the physics of the atmosphere are not yet sufficiently understood to provide conclusive proof, the argument is that such emissions deprive Scandanavians of their right to unpolluted waters.

The five types of interdependence--cognitive, opportuntiy cost, contingency, constraint and deprivation--describe the type of collective situation that exists. A second influence on collective response is the "location" of the interdependence with respect to the relevant group or organization.

By the "location" of an interdependence is meant the functional distance of the basis for collective action from the central interests and decisions of the group or organization in question. It is, in other words, a characterization of the way that the group or organization relates to the situation of interdependence. Ruggie (1975) has identified four typical patterns: external, external manifestation, "commons" and internal.

External interdependence refers to a situation in which the basis of interdependence is outside the normal concerns of the group or organization and affect their members only indirectly. This is the case which is most common when the politicization of issues on the international level has led to the demand for collective production of information. Whether the stituation involves statistics about overfishing, studies of the comparative advantages of different science policies, or innovations in the design of nuclear reactors, the link between the situation of interdependence and the domestic or internal scene is indirect and must be established.

External manifestation of interdependence refers to the situation in which the interdependence involves the extension of internal concerns to the arena outside the group or organization. There are many examples of this, including weather observation, environmental and oceanographic monitoring, civil aviation, telecommunications, and the use of the frequency spectrum. In each instance, the external manifestation is an interdependence on a topic that would be considered an issue at the local level. This kind of interdependence characterizes many different kinds of international organizations (e.g., the Friends of the Earth, the International Chamber of Commerce, the International Council of Scientific Unions). The problem is to establish an international system within which each national or local group becomes compatable with each other. Although the locus of interdependence remains external to the organization or group, the link to its immediate concerns is more direct.

The term "commons" refers to a limited resource readily available to all. In his classic article, Hardin (1968) illustrated the "tragedy of the commons" with a pasture open to all herdsmen. It is rational for each herdsman to try to keep as many cattle as possible on the common pasture. In calculating his own utility in his choice of whether to add another animal, he introduces an inevitable bias. The benefits derived from each animal he adds are his alone; but the costs, and especially the contribution to overgrazing, are shared among all the herdsmen. Each herdsman performs the same calculation and draws the same conclusion--rational from his own point of view--to add as many animals as he can, resulting in the overcrowding of the common pasture. The disasterous consequences of the subsequent overgrazing will, of course, affect each and every herdsman. But even if individual herdsmen recognize the problem and constrain their expansion, the tragedy may still continue its course. Worse, any herdsman so constraining his own herd will place himself at a disadvantage compared with his neighbors, who continue to expand. If overgrazing continues its full course, he is likely to be in a much worse position as conditions become difficult. Only if all herdsmen limit the size to their herds can the disaster be avoided. The situation of interdependence is one of determining property rights for shared, limited resources. Clear examples of international commons interdependencies include the control of the seabed and of the high atmosphere and deep space. Both are subject to extensive international debate and negotiation. In this type of interdependence, the interest is in being able to use the resource in question. The location of the interdependence is external, but the linkage to internal interests is direct and allocative.

Internal interdependence refers to the situation in which internal concerns are repeated in the various groups or organizations involved. Attempts to control the use of nuclear materials, establish environmental pollution standards, and impose livestock feed policies are examples of internal issues that have become linked with those of other nations. The issue is what should be done internally, but it is debated, at least in part, in terms of the consequences of internal measures on others external to the decision and to the subsequent action. A clear example involves the automobile industry. Each of the major automobile producing nations views the auto industry as vital to its industrial structure. Yet the international competition in the industry insures that actions in one country have consequences in other countries. Not only are the consequences in other countries a part of the debate about domestic options, but the actions in foreign countries are sometimes the stimulus for domestic measures. The interdependence is internal: the problem is the domestic industry, and the only actions which can be taken are domestic. But there is a clear recognition of sharing the situation with others.

Two sets of characteristics have been presented to describe the basis for collective action, the type of interdependence and its location with respect to the individual group or organization. The type of interdependence--cognitive, opportunity cost, contingency, constraint or deprivation--suggests the character of the collective situation, and the location of the interdependence -- external, external manifestation, "commons" or internal--suggests the linkage of the issues involved to the usual concerns of the groups and organizations affected.

The overall collective situation described by these two sets of factors is essentially social in nature, not physical or technological. It emerges not so much from the nature of the problems confronted as from the pattern of interaction negotiated by the various actors and groups involved. Each collective situation represents an agreement that one particular configuration and not some other represents the situation. Even when there is such agreement and the basis for collective action exists, it may be seen to be tentative and inherently unstable. It may change as knowledge of the relevant cause and effect relations increases, as configurations of interdependence expand or contract to encompass different sets of actors, or as the opportunities and objectives of relevant groups or organizations change. The basis for collective action is therefore subject to constant renegotiation. Cooperation is not a matter to be decided, but to be cultivated and managed.

<u>Strategy</u>

The previous section, describing the parameters of choice and the type and focus of the interdependence involved, set the stage for the examination of strategies for achieving particular goals. The plan of action selected by an individual, group or organization is assumed to be tailored to achieve the desired goal in the situation in which the person or persons find himself or themselves. In this section, reference will be made only to groups and organizations. But the observations and conclusions about appropriate strategies in different situations could be applied to individuals as well.

Emery and Trist (1965) provided a pioneering work in the description of appropriate strategies for dealing with the kind of uncertainty associated with the tentative and shifting interdependencies presented in the previous section. They used the concept of an open system (in which the entity under study imports certain types of material from its environment, and exports other types back into the environment) as opposed to a closed system (in which all changes are internal to the entity under examination). An important characteristic of an open system is that stability can be attained in a time-independent steady state. Such steady states are quite different from the equilibrium states described in classical physics, although the latter have often been taken as models for biological and social phenomena. Equilibrium states follow the second law of thermodynamics, so that no additional work can be done when equilibrium has been reached. However, the openness to the environment (i.e., the import-export of material) of a system in steady state maintains the capability for work. Without such capability, adaptability (and thereby survival) would be impossible.

In examining such an open system, there are four different types of exchange, or interdependency, which are important: internal interdependence (1), transactional interdependence to the environment (2) or from it (3), and processes by which different parts of the environment become related to one another (4). Emery and Trist call the latter, the environmental interdependencies, "causal texture" and point out that the texture of the environment of an organization is a strong influence on how it organizes its actions.

Emery & Trist (1965, pp. 24-26) describe four different types of causal texture for organizations and the strategies appropriate for each. They may be thought of as "ideal types", and actual situations will differ from the abstract descriptions to some degree. The causal textures increase in complexity in the order they are presented here, and each subsequent type introduces modifications on the conditions in those discussed prior to it.

The simplest kind of external texture is called the "placid randomized environment" by Emery and Trist. It is a context in which the goals and noxiants (goods and bads) are relatively unchanging and randomly distributed. It corresponds to Simon's idea of a surface over which an organism can move which is mostly bare, but with widely scattered bits of food (Simon, 1957) or to Ashby's limiting case of no connection between environmental parts. (Ashby, 1960) It also corresponds to the standard economic conditions of a pure market.

A critical property of organizational response under

random conditions is the lack of a distinction between tactics and strategy. The optimal strategy is the simple tactic of doing one's best on a purely local basis. The best tactic, however, can only be learned by trial and error, and only for particular environmental variances. Units which function well under these conditions are usually small and adaptive, and are not as successful in other types of environments. (Emery & Trist, 1965, pp. 24-25)

The secood type of external texture is called the "placid clustered environment". Goals and noxiants are not random, but interrelated in certain ways, corresponding to Ashby's serial system or to the standard economic conditions of imperfect competition. It means that some parts of the environment take on roles as signs of other parts, or become means-objects in terms of approaching or avoiding.

A new feature of clustered environments is that strategy is distinguishable from tactics, and is important for survival. The success of the organization depends upon what it knows about its environment. To pursue a local goal may lead it into parts of the field fraught with danger, while avoidance of an immediate difficulty may lead it away from potentially rewarding areas. In the clustered environment the relevant objective is that of "optimal location", with some positions being discernable as potentially richer than others. (Emery & Trist, 1965, p.25) Organizations under these conditions tend to be larger and hierarchically organized, with a tendency toward centralized control and coordination.

The next level of causal texturing is called the "disturbed-reactive environment". It corresponds to Ashby's ultra-stable system or to the conditions of oligopolic competetion. There are more than one organization of the same kind. In fact, the existence of several similar organizations is the dominant characteristic of the environmental field. Each organization does not simply have to take account the others when they meet at random, but must also consider that what it knows can also be known by the others as well. The part of the environment to which the organization wishes to move in the long run is also the part of the environment which the others will seek. Knowing this, each will try to improve its chances by hindering the others, and each will know that the others will not only try to do likewise, but also know that each knows this.

If strategy is a matter of selecting the objective--of where one wishes to be at a future time--and tactics a matter of selecting an immediate action from the available repertoire, then there appears in the disturbed-reactive environment an intermediate level of organizational response, the "operation". One must not only make sequential choices, but also choose actions that will draw off the other organizations. An operation consists of a campaign involving a planned series of tactical initiatives, calculated response by others, and counter-actions. The flexibility required encourages a certain decentralization in the organization and also puts a premium on quality and speed of decision at various peripheral points. (Emery & Trist, 1965, pp.25-26)

The final, and most complex, external texture is called the "turbulent field" by Emery and Trist. In this environment, dynamic processes, which create a significant variance for the component organizations, arise from the field itself. There is not only the possibility for goals and noxiants to change and for response on the part of other organizations, but the very "ground" itself is in motion. They describe three factors that contribute to dynamic changes in the field within which the entity finds itself:

1) The response to clustered fields, and the resulting linked sets of organizations, is so pervasive that self-induced processes are introduced into the environment.

2) The increasing interdependence among the units in the environment creates new conditions, especially in terms of restrictions on action as in laws or regulations in society.

3) The increasing reliance on knowledge about goals and noxiants, as well as the interdependencies in the environment, (which is critical for meeting competative challanges) itself stimulates new interactions.

This means an important increase in the uncertainty under which an organization must operate. The consequences which flow from their actions lead off in ways that become increasilgly unpredictable; they do not necessarily become less important with distance, but may at some point increase beyond all expectation. (Emery & Trist, 1965, p. 26)

Emery and Trist (1965, p. 28) point out that virtually all societies respond to such uncertainty about individual direct action by developing values that have overriding significance for all members of the field. Unable to trace out the consequences of their own actions as these amplify and resonate through extended social fields, people have sought rules, sometimes categorical, to provide them with a guide and a ready calculus. So far as effective values emerge, the character of turbulent fields changes dramatically. The relevance of large classes of events no longer has to be sought in the intricate interactions of diverging causal strands, but is given directly by the ethical code. An uncertain, turbulent field is thereby transformed into one of the simpler environments.

There are, however, organizational responses to turbulent fields which are fundamentally different from the hierarchical forms appropriate for other environments. Whereas the disturbed-reactive environment requires one or another form of accomodation between similar organizations which are competing and whose fates are therefore to some degree negatively correlated, turbulent environments require some relationship between dissimilar organizations whose fates are, basically, positively correlated. Thus, relationships that maximize cooperation and which recognize that no one organization can take over the role of other organizations become paramont. They refer to this type of relationship as an "organizational matrix". (Emery & Trist, 1965, p. 29)

As with the emergence of effective values, matrix organizations, if successful, will only help in transforming turbulent fields into the types of environments described as clustered or disturbed-reactive. Though such transformations should achieve a degree of stability for the application of strategy, operations and tactics, the resulting environments would not be identical with the simpler types. The strategic objective in the transformed cases could no longer be stated simply in terms of optimal location or capability. It would have to be formulated in terms of institutionalization. Organizations become institutions through the embodyment of organizational values which relate them to the wider society. (Selznick, 1957) Insofar as institutionalization becomes a prerequisite for stability, the determination of strategic planning will necessitate not only a bias towards goals that are congruent with the organization's own character and aims, but also towards a selection of goal-paths that offer maximum convergence as regards the interests of other parties. (Emery & Trist, 1965. p. 29)

Corresponding to this approach, the important components of a structured response to technological issues would include tactics, operations, strategy and institutionalization. Each serves to define options which "nest" inside the concerns of the next. As the complexity of the environmental texture is "reduced" through appropriate action, effectively simplifying the choices to the level of a less dynamic and less uncertain type of environment, the difficulties confronting the organization or decision maker become easier to deal with. But although the choices are in some ways simplified, there remains the imperative of maintaining the essentially cooperative nature of the effort.

It is clear that as the level of uncertainty and interdependence comes to dominate, in the turbulent field, the appropriate strategies take on a rather new form. The idea that dissimilar organizations must cooperate to accomplish their respective aims as opposed to the traditional concept of similar organizations competing with one another poses quite different problems. Both the resulting matrix organization approach and the adoption of ethical guidelines through institutionalization are substantial deviations from traditional organizational strategies.

Coordination

For a cooperative effort to achieve effective collective action, the individual contributions of the various participants generally must be coordinated. The coordination of collective action can be described in terms of the instruments of integration, the principal decision making character of the key actors, and the functions of the various entities involved.

Ruggie (1975, pp. 570-573) describes four different "instrumentalities" with which national behavior is integrated in what are called "international regimes" (a set of mutual expectations, rules and regulations accepted by a group of states). They are: a common framework to which independent activities can be related, a joint facility seeking to make independent activities commensurate, a common policy giving coherence to separate activities, and centralized direction of separate activities. In the order presented, these may be seen as increasing in the degree of centralized control exercized. For a sense of completeness, therefore, it is possible to add a fifth: the welding of activity into a single agency with traditional hierarchical control. Although these alternative instruments of coordination have been conceived in terms of cooperative efforts on the international level, they can inform us about the ways coordination of response to technological hazards might be achieved at any level of social aggregation.

A common framework attempts to stimulate cooperation through the power of its ideas. To the extent that a common framework actually encompasses the interests of the various entities involved in such a way as to emphasize the commonality of their problems, opportunities and desires, one can expect that simply the definition of a common approach can achieve a degree of coordination. But this is the weakest of the instruments of coordination, and includes no mechanisms to insure that the aims set out in the common policy are in fact followed by those participating. Examples of such frameworks include the Long-Term and Expanded Programme of Oceanic Exploration and Research (LEPOR) and the Global Atmospheric Research Program (GARP). It is fairly common to find problem-oriented research programs that are designed jointly on the international level, but for which the conduct of the contributing research tasks is left to scientists in various participating countries.

A joint facility seeks actively to make the actions of the participating units mesh with each other. It harmonizes definitions and approaches, establishes standards, and attempts to insure the comparability of the activities being conducted independently by the participating entities. Many international organizations are examples of this kind of coordination. The OECD (Organization for Economic Cooperation and Development) functions in this manner in its attempts to harmonize the activities in its member states through the decisions of its governing board. In many respects, the Commission of the European Communities functions as a joint facility. A Community directive is a regulation which has been jointly formulated and which is applied to activities in the member states. Other such organizations include the secretariat of the General Agreement on Trade and Tariffs (GATT) and the International Atomic Energy Agency (IAEA). Each of these organizations has access to some sort of mechanism for insuring or encouraging participating countries to abide by the standards and other findings of the coordinating body.

A common policy integrating separate activity goes further to provide a more precise ordering and scheduling of separate activity. Several functions of the European Community, such as the common agricultural policy, represent this kind of coordination. They are generally more controversial, precisely because they establish an of ordering of the activities among the participating entities. Similar instruments of coordination are the European Space Agency and the International Telecommunications Satelite Consortium (Intelsat).

The coordination of activity through central direction is less common on the international scene, although there are a few examples, such as the control of air traffic through Eurocontrol. More common is the establishment of an agency, with traditional hierarchical control, charged with dealing with a specific issue. Examples of this approach abound. Interpol not only organizes exchange of information among police organizations, but conducts activities of its own. The same can be said of the principal organizations of the United Nations (e.g., UNESCO, WHO, FAO, UNEP) and of organizations like the World Bank and the International Monetary Fund.

These five instruments of coordination--common framework, joint facility, common policy, central direction, and agency--constitute the mechanisms through which cooperative activity is most commonly structured.

A second important area for the description of coordination is the exercise of influence by key individuals. Cox & Jacobson (1973, pp. 11-14) describe typical actors involved in the normal affairs of international organizations and characterize the ways they exercise influence. The provide the following categories: 1) representatives of national governments, 2) representatives of national and international private associations (including interest groups and commercial enterprises), 3) executive heads of organizations, 4) high officials and other members of the bureaucracy of such organizations, 5) individuals who serve in their own capacity formally or informally as advisors, 6) representatives of other international organizations, and 7) employees of the mass media. This set of categories represents, of course, a rather more differentiated view of an organization, its social environment, and the potential exchanges than has been presented up to now. Such differentiation, however, is in no way inconsistent with previous concepts, and can help to provide realistic characterizations of conditions as they are encountered in actuality. Not all categories of actors will be active in every case, nor is is likely that every actor participating in a given organization will be involved in every decision. Nevertheless, individual participants fall in one or more of these categories, and the patterns typifying their influence on the outcomes of organizational decisions in an important element in the characterization of coordination.

Influence can be exercized in many ways, but Cox & Jacobson (1973) list four as particularly significant. In each decision some one or group must take the initiative, and hence may be called the "initiator" of the proposed action. The actual initiators may not be those who appear to do so in the formal process. For example, in organizations which include representative voting bodies, the members of the administrative staff often prepare documents which are submitted by delegates with no formal connection to the authors. The identification of initiators is therefore often a matter of critical inquiry, not just of simple observation. Some individuals or groups, perhaps because of their crutial location in the channels of communication or of their control over political and financial resources, have the ability to block iniatives, even though they may have no particular advantage in terms of initiation. Hence they may be called "vetoers". This term should not be understood in the formal, legal sense, but in a practical, functional sense of having the power to prevent a decision or its implementation by whatever means. There are also individuals whose views, either known or summized, must be taken into account, perhaps because of their control of resources or of their formal authority. They can be called "controllers". Finally, there are those who serve to bridge the gaps between different factions and who develop compromise or build consensus. They can be called "brokers".

Each of these different kinds of influence can be important. Their relative importance in a particular case depends upon the overall situation, the way the system in question functions, and the patterns of interaction that are typical of it.

Two sets of factors have been described as important for coordination: the instruments whereby coordination is structued, and the manner in which individuals exert influence. The final element to be presented here is the functional nature of the organizational units involved.

The classification of the nature or function of the entities under examination is an attempt to capture in a set of simple categories the important aspects of the way that the attributes of the organization as such affect the way that people belonging to it are likely to view problems and their possible solution. Four different functional orientations are considered important in this respect: informational, managerial, executive, and special interest.

The principal function of informational organizations is the production of knowledge. Of course, the most obvious examples are research units, which may be found in commercial enterprises, governmental agencies, and universities as well as in international cooperative programs. But purely informational functions can also be found in control programs, as for example in the United Nations program for safeguarding nuclear materials. The function of the "investigatory" unit is simply discovering the diversion of materials and reporting it to the security council, and includes no responsibility for preventing such diversion or imposing sanctions should it occur. Its function is purely informational.

The managerial function is one of overseeing the conduct of affairs, usually in terms of previously determined goals, methods, and resources. The principal characteristic of this function is that it includes allocative choices in directing the application of resources to achieve the expressed goals. Examples include Intelsat, which manages operational satellites, or Eurocontrol, which manages commercial and private traffic in the upper airspace.

The executive function is that of exerting control over the course of affairs through application of ethical or other normative guidelines together with the imposition of sanctions of various sorts for violations. Less common, but also possible, is the offering of incentives to encourage adherence to the relevant guidelines. In this sence, the executive function refers not only to the chief executive officer and his staff in a commercial enterprise or the head of government and cabinet in public affairs, but also to the judiciary, and sometimes, to the members of the professions. Of course, the courts are the most obvious arena for the exercise of ethical control. But as the above indicates, ethical and normative concerns influence the course of events in a variety of less formal and indirect ways as well.

The functional orientation of the organization or unit within an organization can often influence the way members of that entity conceive problems and their possible solution. In this way the function of the unit is important, not only in terms of identification of the of the possible actions, but also in terms of the potential contribution from various sources. It can play an important role in the negotiation process.

The four major factors that should be included in considering coordination are therefore the instruments of coordination, the role of key individuals, the exercise of influence and the functional orientation of the organizational units involved. With information in each of these four categories, a picture of the potential possibilities for coordination can readily be built up.

Summary and Conclusions

The topic addressed in this chapter is the collective response to technological issues. It is a difficult topic for several reasons: because of the large number of conceivable responses to issues involving large-scale technologies, because of the variety of concepts that might be used to describe such actions, because of the paucity of research linking collective response to technological hazards with the formal analysis of such hazards. One result is that this chapter, compared to the previous one on individual reaction, must rely on relatively preliminary and speculative conclusions rather than on the results of empirical research designed to clarify issues related to technological conflicts. This is not to say that there is no research about collective response to technological issues. To the contrary, there is so much analysis that might conceivably be applicable that it is difficult to select items that are appropriate for the specific topics of this chapter. The difficulty is matching available analyses to the focus of this study.

For the most part, there is a large and significant gap in the research that might constitute the basis for this chapter. Considerable abstract analysis has been done describing various responses to technologies as well as examining the respective advantages and disadvantages of these responses. At another level, there have been innumerable case studies which examine particular sets of events and decisions about the development or control of technologies. But there is very little work attempting to bring the structural analysis in the former together with empirical data in the latter in ways which apply to the management of technological hazards per se. Such an approach is admittedly difficult, but the scarcity of such research makes the task of this chapter much more difficult.

Descriptive categories are developed in this chapter which enable the characterization of the principal attributes of collective response to technological hazards. Although they exhibit a logical progression roughly corresponding to a stylized model of decision making in general (setting parameters, determining interdependencies, selecting strategies, effecting coordinated response), they cannot be said to be predictive in that they do not suggest the response that might be encountered under given conditions. But even though the limitations of such an approach are clear, the ordering of the attributes of the phenomenon under examination can be quite useful. The four sets of categories provide a plausible framework for the description of collective response to technological issues.

A few conclusions may be drawn on the basis of these categories, even if not directly supported by expirical research. For example, it is clear that the technical elements of technological issues are only one of several kinds of influences on collective behavior. Just as the economist finds human interaction to be costly, requiring the concept of transaction costs as well as the ordinary costs of production, so we find the elements of interaction join the hazards and their consequences as important influences on collective action. Thus, it may be concluded that the aims and interests of the relevant groups or organizations influence the type of interdependency upon which collective response can be based, just as does the nature of the technical cause and effect. The chain of cause and effect constitutes the original impetus, but the interests of the respective groups are also necessary to define the exact nature of the interdependency among the affected groups.

This observation can even be extended. The section on the parameters of choice casts the boundaries of consideration in terms of goals and normative preferences as well as cause and effect. This is unavoidable in dealing with collective action. Groups (and organizations) exist with purposes and aims (see Chapter 4), and if the actions of different groups are to be coordinated, their respective aims must be considered the starting point. It should be no surprise that as the level of focus shifts from the individual to the group the principal motivations for action shift from the psychological to the sociological or political.

The economic theory of clubs, described briefly in the introduction to this chapter, suggests that collective action will be at a level which is too low (corresponding to free ridership) or too high (as with zealotry). This chapter, however, focusses more on the nature of the collective action and how to effect it than on the level of activity and whether or not it is appropriate. Thus, the key factors of those examined may be said to be the identification of interdependence and the effecting of coordination. In Chapter 4 it was stated that the three principal ways of obtaining support for one's program of action are the exercise of authority (direction of others), exchange (on a quid pro quo basis) and persuasion (agreement on the need for action). The identification of the interdependence upon which collective action can be justified is therefore of utmost importance for a program of action in a pluralistic setting. Even when support is gained through exercise of authority or through exchange, the existence of some sort of interdependence can provide an essential justification for action. This chapter provides a categorization of several different types of interdependence which may not only assist in recognition of the nature of the interdependence in a given situation but also suggest the best way of effecting coordination.

The section on coordination also extends the analysis into more detail concerning the possible actions on the part of groups and the individuals who make them up. The set of categories presented there also provides the capability of dealing with the greater complexity encountered in actual situations, and can be combined with the conceptual framework in Chapter 4 to give a fairly detailed picture of how one might proceed in a given situation. This could easily include strategies for interacting with other individuals or groups in situations of conflict. But as mentioned earlier, the ideas presented in this chapter need to be applied to actual situations in order to determine their true usefulness. Such empirical applications are not available at this time.

One final point should be made. Since collective action is dependent upon the respective aims and interests of the groups involved, one may expect that unanimity would be the exception rather than the rule. Further, while some individuals or groups might strive to increase the stability in a given situation, others might find it in their advantage to disrupt that stability. Thus, it may be quite important to recognize situations which call for cooperation between dissimilar organizations or groups whose aims are somehow positively correlated (as opposed to the "traditional" situation in which the aims of similar entities are negatively correlated). Again, the recognition of interdependencies is the key, followed by the securing of support and effecting of coordinated action. It is not inappropriate to repeat that although the technical causes provide the ultimate source of such interdependencies, they are equally dependent upon the "social" factors associated with the respective groups.

Although the introduction of social and political factors makes the choice of appropriate response to technological issues even more complex, their inclusion is essential for a realistic treatment. An approach to the management of technological hazards which excludes them would be incomplete.

7 SYSTEMIC ATTRIBUTES OF MODERN TECHNOLOGICAL HAZARDS

Although individual reaction and collective response contribute a great deal towards a basis for the rational and intentional management of technological hazards, there still remains another important area of influence. This chapter addresses an additional level beyond those of the individual and the group or organization--that of cultural or societal influence.

It is possible to distinguish between technical and cultural attributes of technologies and their hazards. Technical attributes, which may be thought of as the patterns of physical cause and effect, are usually relatively stable: they show little change over time or from place to place. But the cultural attributes, essentially the organizational and behavioral patterns associated with the physical artifacts and their hazards, often exhibit considerable variation. Virtually interchangable physical configurations are often handled quite differently in various places, for reasons that might best be called historical accident. The respective courses of events simply developed differently, for reasons that may or may not be clear. Furthermore, social and organizational patterns change over time. Like language, where new words emerge at fairly regular intervals, social and organizational patterns evolve, with new mechanisms emerging tailored to new problems or to new solutions for existing problems.

Many influences on organizational and social patterns

associated with technologies and their hazards derive from the interactions among relevant groups and organizations as described in the previous chapter. However, there are other influences on these patterns more properly called cultural or societal. It is these latter which are the subject of this chapter.

In their book Risk and Culture , Douglas and Wildavsky (1982, p. 8) point out that the selection of particular kinds of danger for public attention and action is a cultural phenomenon.

> The choice of risks and the choice of how to live are taken together. Each form of social life has its own typical risk portfolio. Common values lead to common fears (and, by implication, to a common agreement not to fear other things). There is no gap between perception and reality and no correct description of the right behavior, at least not in advance. The real dangers are not known until afterward (there always being alternative hypotheses). In the meantime, acting in the present to ward off future dangers, each social arrangement elevates some risks to a high peak and depresses others below sight. This culural bias is integral to social organization.

Thus, the risks we face are chosen in the same package as are our social institutions. This is the topic of this chapter; not the influence of industry structure or economic systems, but the deep seated predeterminations that derive from the cultural patterns of society as a whole.

It might be possible to address the cultural influences on the management of technological hazards by examining the overall pattern of institutional arrangements, to examine the more or less "visible" consequences of culture. However, an alternative, the examination of the deeper forces which lead to such institutional arrangements is attempted in this chapter.

Description of deep cultural influences is an ambitious task, and one that requires severe restriction if it is to remain manageable within the limitations of this chapter. No attempt will be made to describe all the social and organizational patterns in modern industrial society. Rather the attempt will be to isolate those cultural patterns which have the greatest impact on the management of technological hazards. The presentation may at first appear disjointed and arbitrary. However, each of the elements discussed contributes to a systematic and consistent influence on determinations about technologies and their hazards. It is the cumulative influence, derived from the predeterminations of the industrial, economic and political systems within which technologies are developed and used, to which the term "systemic attributes" refers.

Four different themes are brought together in this chapter. The first section examines the assumptions and preferences of two extreme positions with respect to technological development: ecocentrism and technocentrism. Both of these positions emphasize the need for careful analysis of human activity. The second section looks at such analysis in terms of the politicization of science and technology. Next, the dependent participation of the majority of the populace and the resulting appeal for democratic decentralization, a phenomenon which appears in many technological conflicts, is examined. Finally, the personal experience of cultural patterns by the individual, and the ways this experience influences his or her preferences and strategies for action is addressed. In this manner, the systemic influences treated in this chapter are tied to the conceptual framework presented in Chapter 4.

Environmentalism

An important influence on both the reaction to technologies and their hazards and on the response to them may be found in the set of principles embodied in environmentalism. Perhaps the best way to clarify this is to contrast the two major conceptual positions concerning technological development. (O'Riordan, 1981)

One view is that there exists a natural order, which continued in a delicate but perfect balance until man intervened. The actions of man have disrupted this balance and led to degeneration of the natural order, according to this view. Further ignorant and presumptive human imposition on the natural order will ultimately lead to the destruction of mankind. (McConnel, 1965). This view has been called "ecocentrism" or the "ecocentric mode". (O'Riordan, 1981, p. 1)

The second position calls for the application of rational scientific and managerial techniques by a professional elite, who regard the natural environment as "neutral stuff" from which man may shape his destiny. (Hays, 1959) This position has been called "technocentrism" or the "technocentric mode". (O'Riordan, 1981, p. 1)

These two perspectives differ not only in their views of nature, but also with respect to their underpinnings in terms of morality. Ecocentrism advocates respect, humility, responsibility and care; it opposes bigness (especially in cities), and espouses a code of behavior based on permanence and stability. Technocentrism, in contrast, assumes an almost unlimited capacity on the part of man to understand and control events to suit human purpose, extending even to the prediction and manipulation of changes in value systems and behavior. Ecocentrism is concerned with both the ends and means to be met and the ways these aims are to be carried out. Technocentrism focusses on the means, especially on managerial principles. Ecocentrists have, until quite recently, generally avoided political action and have limited their activity to rhetoric and intellectual persuasion. Technocentrists are often urban-dwelling elites who thrive on sophisticated communications and tend to be politically influential. (O'Riordan, 1981, p. 1)

Of course, as is the case with many characterizations of groups who are politically active, such a crude splitting into two diametrically opposed extremes is an exageration. In reality the positions of ecocentrism and technocentrism are blurred. It is probably true that environmentalists favor ideas drawn from each ideological position, depending on the particular issue at hand. Nevertheless, each of these sets of ideas have influenced the development of modern environmental activities.

O'Riordan (1979, pp. 10-11) summarizes the effects of ecocentrism on the environmental movement as follows:

> First, it provides a natural moralty -- a set of rules for man's behavior based upon the limits and obligations imposed by natural ecosystems. Taken to its extreme, this natural morality displaces the humanist morality that is intrinsically derived through man's cultural institutions. But even in its more moderate form, ecocentrism provides checks to the headlong pursuit of "progress" which, by and large, is the object of the technocratic mode.
>
> Second, it talks of limits (blurred perhaps, but recognizable) of energy flows or productive capacity, and of the costs of organization and system maintenance.
>
> Third, it talks in ecosystem metaphors of permanence and stability , diversity, creativity, homeostasis and the protection of

> options. These have important policy ramifications for the preservation of unique ecological and/or cultural habitats and lifestyles.
>
> Fourth, it raises questions about ends and means , particularly the nature of democracy, participation, communication among groups holding conflicting yet legitimate convictions, the distribution of political power and economic wealth, and the importance of personal responsibility. These questions have profound political overtones, few of which are fully understood even by politicians themselves.
>
> Finally, ecocentrism preaches the virtues of self-reliance and self-sufficiency. Here again, the political ramifications of this trend are not properly understood. Anarchist environmentalists aim to develop decentralized institutional arrangements to facilitate flexible and adaptive responses to changing circumstances by individuals and communities alike, and to avoid the vulnerability of dependence upon trade, food supplies, large corporations, or key occupational groups.

O'Riordan (1979, p. 19) then goes on to describe the impact of the technocentric position on environmental policy making in the following terms:

> First, its optimism over the successful manipulation of techniques to extract and allocate resources -- an optimism shared by most policy makers who bask in the reflected glory of technological success.
>
> Second, its determination to be " value free " in advice and analysis, leaving the "tough" judgements to a political arena that is already shaped by their advice.
>
> Third, its disavowal of widespread public participation , especially the input of lay opinion, a philosophy much admired by politicians equally intent on preserving their rightful role in acting authoritatively on behalf of the public. This makes it difficult to insure that minority views, or "nonquantifiable" factors such as preferences for particular kinds of lifestyles, are adequately taken into account.
>
> Finally, its disquieting fallability , the constant evidence of error and misinterpretation and of hunches that do not

> quite pay off. Fallability is tolerable when those in important positions ignor its existence. While the Achilles heel of fallability is now the target of informed scientific opinion, the result can be endless technical debate, prevarication and exasperation, all adding to the uncertainties ever present in environmental policy making. As the credibility of the technocrat is increasingly questioned, more expensive and more careful scrutiny strips off the veil of optimism, revealing a hesitancy that opens up wide avenues for political intervention and public participation.

The most important point, made quite clearly by O'Riordan, is that both positions vis a vis technological development (i.e., open opposition versus rational or planned implementation) lead to critical scientific review of programs and projects. This, in turn, has led to the politicization of science and technology, which is examined in more detail in the next section.

The Politicization of Science and Technology

It is often said that science and technology have become politicized. But it is seldom specifed what is meant by "politicization". This refers, in general terms, to the feeling that the results of scientific research are used to support or justify actions undertaken for essentially political reasons on the one hand, and that science and technology are controlled by political decision makers on the other. But sometimes the term is used merely to indicate the fact that science and technology have become more controversial, more the subject of public debate and discussion.

In this process, however, it is important to distinguish the politicization of science from the politicization of technology. The terms "science" and "technology" are so often found in combination that it sometimes seems that they are intellectual siamese twins, and anything affecting the one must necessarily affect the other. Of course, it is true that a great many practical technologies can be traced directly to the results of scientific research. It is equally true that scientists have used this fact to "justify" the allocation of public funds to research activities. (Bermant, 1981) But both the conduct of science and the application of technology have ties to other elements in modern society, as well. Whether the strength of the connection between science and technology is fact or myth is less important to this discussion than the similarities and differences in the forces that have "politicized" each.

There are three important themes that characterize the politicization of science and of technology. The first is that some research activities and some technological developments are feasible only with the commitment of government. This is usually, though not always, because of the resources required (not only money, but sometimes scarce materials, equipment or expertise). Accompanying the necessity of government support has been the shifting of decisions to higher and higher levels within the heierarchy of government. In some circumstances, this results in the treatment of science or technology as just one more element in the political calculus. The second broad theme in the politicization of science and technology involves the enactment, in nearly all industrialized countries, of legislation aimed at the control of scientific and technological hazards and abuses. Finally, the public has become much more aware of certain scientific and technological controversies. Divergent views spread, conflicts rage, and citizens band together to insure that their voices are heard in the relevant arenas of decision making. Sometimes citizens engage in actions intended to hinder the course of events or even to block developments to which they object. Each of these broad topics--the involvement of the highest levels of government, the establishment of legislative and administrative controls, and public discussion and citizen action--reflect an important aspect of the politicization of science and technology.

It is tempting to use the developments in the nuclear industry to illustrate the necessary involvement of government in the performance of research and the development of technology. To be sure, government action was unavoidable in the nuclear industry, not only because of the financial and resource requirements, but also because of the overriding importance of nuclear technologies for military interests and national security. But nuclear power is in many respects a special case which involves several factors that do not apply to technologies in general. It is perhaps better to focus on technologies reflecting more "common" developments.

Many other technological developments have also relied on government support. Most energy production technologies, e.g., fusion, solar collection or conversion, synthetic fuels, appear to require public support of research and development efforts. (That public funds are "required" is perhaps questionable. But public funds are devoted to such projects with the justification that progress in the field would otherwise be too slow or might even be prematurely abandoned.)

It is not uncommon to view the general support of technological development as one of the primary functions of government. It is quite typical to find government support for industries that are felt to be vital in some way to the health and stability of the economy in general. Thus, governments tend to pay considerable attention to developments in industries like steel, shipbuilding, automobiles, machine tools, computers, and aircraft. It is common to find public funds devoted to research and development of new and improved technologies, especially in these industries.

The importance of such support activities in modern government is evidenced by the manner in which it is organized. In the Federal Republic of Germany, for example, it has been accorded cabinet status: the Federal Ministry of Research and Technology (the German initials are BMFT). Research is also conducted or sponsored by other ministries, but it is the primary function of the BMFT.

In general, a decision may be said to be more political

the higher the level of the official involved in the deliberation. This follows from a basic property of hierarchical organization: decisions requiring consideration of two or more areas of responsibility (especially when trade-offs between or among them are required) tend to rise to that level of responsibility which encompasses all the areas affected. Decisions are usually shifted upwards because it proved impossible to resolve the competing considerations at a lower level. Thus, it may be assumed that the higher one goes in a hierarchy, the greater the number of considerations will be included in the deliberations. The fact that the Minister of Research and Technology sits on the Federal cabinet suggests that at least some of his decisions about research and technology will reflect consideration of wider interests. This contributes to the view that, especially during periods of low economic growth or stagnation, expenditures of funds for research might reduce the ability to create new jobs or to provide other vital public services. Thus, one can claim that government support of research and technology is today just one more of the many concerns in the polical arena, and is subject to essentially the same political influence as any other concern in that arena.

The second major attribute of the politicization of science and technology may be described in terms of attempts to control the negative consequences of research and applications of technology through legislation and administrative regulations. Of course, the control of hazards is not new. The industrial societies of the West have seen three great waves of government intervention: during the period of industrial expansion in the late 19th century, following the Great Depression in the early 1930s, and following the economic boom of the 1950s and 1960s.

(Business Week, 1979)

The first wave of government intervention, in the late 19th century, instituted fairly simple controls on abuses about which a reasonable consensus had developed: monopoly power, tainted food, child labor. In general, these measures prohibited such activities and gave government the power to punish offenders. The second wave of intervention, following the Great Depression, not only extended prohibitions to new abuses, but added two new dimensions of control. Provisions were established for the manipulation of the economy so as to avoid the excesses of over investment and excess capacity, and for the exercise of preventative actions as well as punishment afterwards. Government interventions could no longer be called simple, especially in so far as they attempted to exercise control over the economy as a whole. Direct intervention into specific industries, in contrast to monetary or fiscal policy, was generally assigned an organization responsible only for the industrial branch it was to control. Thus, government regulations generally reflected motivations not only to exert influence over the industry, but also to insure its long run survival and growth. The most recent set of government interventions, however, introduced a radically different approach. Following nearly thirty years of sustained economic expansion and the revolution in social outlook and mores that accompanied it (see the discussion of the post-materialist orientation in Chapter 5), these most recent interventions reflect a shift of focus from away from the instigators of abuse or inappropriate conduct to those who suffer the abuse and bear the conseqeunces. The success of the Environment Agency in Japan, the Environmental Protection Agency in the United States, or the Umweltbundesamt in the Federal Republic of Germany in reducing harmful factory emissions does not require the survival of the factory doing the emitting. The implicit relationship between the "new" regulatory entities, whose constituency is made up as much of those suffering the conseqeunces as of those associated with production, and the firms to be controlled will in all likelihood insure that a high level of controversy is maintained.

Just as institutionalizing decision making about research and technology at the highest levels of government contributes to the politicization of science and technology, so does the establishment of environmental, occupational and consumer protection regulations ostensibly based on scientific and technical analysis of the effects of relevant activities. Both have tended to increase the visibility of the issues involved, thereby increasing the potential for political response.

The final aspect of the politicization of science and technology to be addressed here is the emergence of both the conduct of research and the use of technology as issues of public conflict in their own right. Nelkin (1979, pp. 11-13) groups such contrroversies into four categories which typify the diversity of concerns in protests against science and technology: 1) questions of equity, 2) fears of hazards, 3) constraints on freedom of choice, and 4) infringements on traditional values.

Questions of equity arise when citizens in a community realize that they must bear the costs of a project that will benefit a different or a much broader constituency. Airports, power stations and industrial facilities are typically the subjects of such controversies. Fears about the potential health or environmental consequences make up the second category.

Sometimes it seems as if every newspaper carries a new warning about another "invisible" hazard like radiation, cyclemates, or sugar, to name a few recent examples. Such warnings are usually about previously unknown consequences of familiar things, or about complicated effects which can be detected and understood only with complex measurement and analysis. Technological advance has increased our ability to detect small amounts of substances and to estimate infintesimal hazards. But remaining uncertainties leave considerable room for conflicting interpretations. The resulting disputes among scientists contribute even more confusion to already difficult controversies. The third kind of controversy involves individual freedom of choice versus government regulation. For example, government bans cyclemates, laetril and other substances it judges dangerous to the public. At the same time it imposes floridation or clorination of water supplies, universal vaccination, or the use of seatbelts in autos. In some cases, citizens begin to question the government's unilateral right to impose its judgment on what the individual may or may not do, especially when the scientific and technical information upon which the judgment is based remains open to contradictory interpretation. Finally, conflict occurs when science or technology is perceived to flaunt traditional values. Contoversies over research procedures (e.g., the conduct of recombinant DNA experiments, or procedures to protect the rights of human subjects) or over the overall impact of science (for example, that it changes the normal state of nature, alters the genetic structure of mankind, or eliminates free will and self-determination) are typical of this kind of protest.

Technical and scientific information play key roles in

such controversies. Access to knowledge and the resulting ability to question the data used to legitimize decisions is a vital capability in these conflicts. Scientists were the first to warn the public about the possible risk accompanying recombinant DNA research, the first to call public attention to the questionable wisdom of developing nuclear power before solving the problem of radiation disposal. (Nelkin, 1979, p. 15) As technical controversies develop, scientists are called upon to buttress political positions with the authority of their expertise. The willingness of scientists to expose technical uncertainties and to lend their expertise to various groups constitutes an important political development.

When expertise is available to both sides of a technical controversy, it is more likely that attention will be called to technical uncertainties, which in turn leads to further polarization. The mere existence of conflicting technical expertise leads to controversy. Often the fact that experts disagree is more important than the substance of their disagreement. After hearing 120 scientists argue over nuclear safety, for example, the California State Legislature concluded that the issues were not, in the end, resolvable by expertise: "The questions involved require value judgments and the voter is no less equipped to make such judgments than the most brilliant Nobel Laureate." (Nelkin, 1979, pp. 16-17) The participation of scientific and technical experts in such disputes can lad directly to demands for greater public involvement in technical decisions.

Each of these developments--the involvement of government in the support of scientific research and the development of technologies, the increase in interventions intended to control hazards and abuses, and the use of scientific and technical expertise to legitimize positions in tehnological contrversies--have contributed to the politicization of science and technology. Scientific research and technological development are no longer viewed simply as positive and unquestioned. Sometimes, at least, they are themselves the subject of challange and controversy. This general attitude constitutes an important aspect of the context of the management of technological hazards.

The New Setting of Social Conflict

Alaine Touraine, a French sociologist, has developed a characterization of modern industrial society which sheds considerable light on the context within which technological hazards must be seen. Touraine (1979, p. 3) believes that a new type of society is being formed:

> These new societies can be labelled post-industrial to stress how different they are from the industrial societies that preceded them, although--in both capitalist and socialist nations--they retain some characteristics of these earlier societies. They may also be called technocratic because of the power that dominates them. Or they can be called programmed societies to define them according to the nature of their production methods and economic organization.

Touraine points out that although economic growth is even more important in this new type of society than for those which preceded it, exactly this emphasis on growth makes it impossible for economic mechanisms to dominate social organization and activity as it did before (Touraine, 1979, pp. 5-6):

> Growth results from a whole complex of social factors, not just from the accumulation of capital. Nowadays, it depends much more directly than ever before on knowledge, and hence on the capacity of society to call forth creativity. All the domains of social life--education, consumption, information, etc.--are being more and more integrated into what used to be called production factors. This is true of scientific and technical research, professional training, the ability to program change and regulate its elements, the management of organizations with multiple social relationships, and the communication of attitudes that favor the mobilization and continued transformation of all these production factors.

It is much more accurate, Touraine feels, to think of growth as resulting from the overall political process rather than simply as dependent on economic activities operating almost independent of social control.

Touraine claims that in modern society the common forms of social domination, the patterns which determine the nature of social conflict over the long run, have been altered. Although we continue to speak of "economic exploitation", we refer to something much more complicated than the determination of an economic exchange by only one of the parties involved. According to Touraine (1979, pp. 7-8), social domination in modern industrial societies takes three forms: social integration, cultural manipulation, and political aggresiveness.

By "social integration" is meant the phenomenon whereby the production process imposes a life-style that matches its objectives and power system. The individual is forced to participate in social organizations which further the aims of production, not only at work but also with respect to consumption and education. "Cultural manipulation" refers to the fact that the conditions which are necessary for growth are not restricted to the realm of production, but influence is equally required over needs and attitudes. Education is no longer the responsibility of the family, or even the school. The individual is expected to continue his education throughout his life, and the mechanisms whereby knowledge is dispersed through the "mass media" offer considerable influence to those who sit at the decision centers of society. "Political agressiveness" characterizes the activities of modern organizations. Modern industrial society is made up of interlocking agencies, dominated by great politico-economic organizations oriented toward their internal functioning as well as their external relations.

In this setting, Touraine finds it more accurate to describe the essential character of the patterns of social conflict in terms of alienation rather than exploitation. Alienation defines a social relationship; exploitation an economic one. The major forms of domination in modern industrial society--social integration, cultural manipulation, and political agressiveness--combine to create a condition of "dependent participation". The individual has two choices: to participate on terms dictated to him, or to "drop out" and lose his ability to play a role within the society. The individual is pressured from every side, by every element of social activity, to assume his role and contribute his bit. People are not alienated because their needs are abused by impersonal interactions, by routinized assembly lines, by urban congestion or by the incursions of the mass media. Alienation comes, according to Touraine, when the individual feels his only relationship to the social and cultural directions of his society to be one accorded to him by others because it is compatible with their continued dominance.

> Ours is a society of alienation, not because it reduces people to misery or because it imposes police restriction, but because it seduces, manipulates, and enforces conformism.
> There are new social conflicts peculiar to the society we observe being formed. Rather than simply a conflict between capital and labor, the new conflict is between the structures of economic and political decision making and those who are reduced to dependent participation. (Touraine, 1979, p. 9)

The nature of these new social conflicts are more social and cultural than economic. Those attempting to counter the domination of dependent participation appeal to widely accepted principles of fairness and democracy to decry the imposition of special interests on the populace as a whole. They resist changes that are not subject to the control of all the people.

> Long enervated by satisfaction with its material success, society is rejecting, not technical progress and economic growth, but its own subjection to a power which boasts of being impersonal and rational, a power which encourages the idea that it is merely a conglomerate of all the demands made by change and progress. (Touraine, 1979, p. 11)

One of the most important determinants of general reactions to technological hazards, perhaps the most important, may be not the hazard itself, but the wider system whereby the hazard and the technology producing it are controlled. Technological hazards can be viewed as yet another aspect of the pattern of dominance in modern industrial society, another instance where the individual is forced to participate whether he wants to or not. Thus, the response to a particular hazard may reflect more a strategy for dealing with the overall control system than the particularities of that hazard alone.

The Influence of Personal Experience on Perception and Stragety

The influence of experience on one's view of the world and on preferred strategies of action can be examined with a concept which Thompson (1980a) calls a theory of political culture. It traces its anticedents to Ruth Benedict's seminal book in anthropology: Patterns of Culture. The basic idea is that culture is neither some kind of rulebook for behavior which is passed, unchanged, from generation to generation nor a simple rationalization of whatever behavior happens to occur. Rather culture is plastic, and although it can be pushed this way or that, it cannot be pushed just anywhere. Furthermore, the conscious attempt to shift culture in a particular fashion requires considerable social effort. Thus, the basic idea is that there are only a few cultural biases which can be shared with others in any given social configuration. Only when a cultural bias is shared is it viable, or stable. The stabilization of cultural bias can be recognized through the emergence of a shared cosmology, or world view. Those with the same cultural bias share the same cosmology.

These cosmologies are made up of beliefs and convictions about the nature of the world which provide the basis for judgment. Each cosmology, since it is associated with a particular bias concerning the basis for judgment, also leads to what may be called a preferred strategy for action. The combination of judgmental bias and preferred strategy found in this political theory of culture constitutes a potentially valuable tool for the description of contextual influences on individual and group response to technological hazards.

Finally, each stable cultural bias (including both the cosmology and preferred strategy) derives from the pattern of experience of the individual. (Thompson calls this pattern the "social context" of the individual. But that term has already been used with a different meaning in this study, so that the phrase "pattern of experience" will be retained.) The pattern of individual experience can be described with just three variables: the degree to which the individual is bound by personal agreements or pacts with other individuals (grid), the degree to which the individual is subject to the constraints of group or organization membership (group), and the degree to which the individual is able to use such pacts or membership to carry out his own wishes as opposed to submitting to the dictates of others (manipulation). (Thompson, 1980c) By examining the individual's "position" in a three dimensional array of the grid, group and manipulation variables, a prediction may be derived of the cosmology, and therefore, of the preferred strategy for action the individual is likely to have.

The two dimensions of grid and group are the key to the description of the pattern of experience associated with a particular cultural bias. Although the degree to which the individual is able to manipulate rather than submit is important in reducing the number of possible stable cosmologies, the interactions of grid and group provide the essential distinction of interest in examining influence over decisions about technological hazards. The simplified model presented here concentrates on these two variables, and makes only limited reference to the manipulation variable. More complete descriptions may be found in Thompson (1980a; 1980c).

The dimensions of grid and group may be characterized in a

two dimensional field as in Figure 7.1. Positive grid, found at the top of the figure, is characterized by a situation which is hierarchical. Interactions with other individuals are carefully based on the respective positions of each person in some relevant hierarchy. A postal clerk, for example, interacts much differently with his fellow clerks than with the head of the postal service. Such interpretations of position within hierarchies and the resulting behavior are usually based on past experience. Thus, a policeman often treats the shaggy occupants of a rattle-trap car differently than the "buergerliche" passengers of a shiny Mercedes, a fact that urban guerillas have duely noted.

Moving to negative grid, at the bottom of the figure, we find a region characterized as "egalitarian". In this region, the status of the individual is not differentiated. Each individual is felt to have the same chance and the same opportunity for success. "Let the best man win" and "all men are brothers" are typical views associated with negative grid.

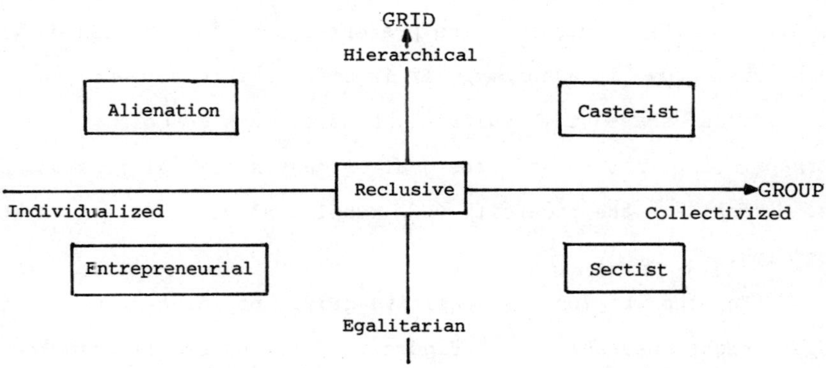

Figure 7.1 Stable Conjunctions of Patterns of Experience and Cosmologies

The group dimension is fairly self-explanatory. In positive group, to the right side of the figure, membership in groups is quite important. In the social configurations associated with positive grid, membership in groups gives the individual both his sense of purpose and the support he needs to deal with a hostile environment. As one moves to the left, to the region of negative group, the picture changes considerably. A person experiencing negative group receives neither sense of purpose nor support from group membership. To the contrary, the individual is either excluded from such groups, or his behavior is such to preclude membership in them.

In the exact middle of Figure 7.1 is a region associated with zero-grid and zero-group. In this region, the individual is neither included in nor excluded from group membership, but rather chooses indifference. Similarly, there is independence from personal pacts. Thus, it is easy to understand that the cosmology associated with this position is called "reclusive". A hermit, or reclusive, strives to avoid both personal and group involvement. He wishes neither to manipulate others nor to submit to their wishes. The preferred strategy associated with this cosmology is autonomy. It is difficult to achieve, and few individuals share the cultural bias of this cosmology. It is, perhaps for that reason, the goal of many spiritual systems, and is, despite the scarcity of people sharing it, a stable cosmology.

In the region of positive-grid, positive-group in the upper right quadrant of the figure is a second stable cosmology. It is labelled "cast-ist" because its fundamental distinguishing characteristic is the strict differentiation of groups according to their respective status. Both the status of the group in terms of some relative hierarchy, and membership in the group are carefully defined. Presentation of oneself as belonging to a group when unwarranted is viewed by this cosmology as a strong breach of social etiquette. Such violations of social propriety are treated harshly, sometimes leading to expulsion from the entire system of group support. In this complicated system of membership and satus relations, leadership functions on many levels, each highly specialized. This provides a clear, multi-layered career pattern within which leaders tend to be quite mobile. Ritual and standardized procedures are valued, because they embody the prescriptions of hierarchy and group specialization. Membership is by recruitment, and invitations are extended only to those who have demonstrated themselves to be acceptable. The caste-ist cosmology encourages a high degree of manipulation, but only in terms of the maintenance and protection of status groups. The preferred strategy associated with the caste-ist cosmology is called the manipulative collectivist strategy.

The region of negative-grid, postitiv-group in the lower right quadrant of the figure is occupied by a stable cosmology called "sectist". In this region personal experience is of egalitarian groups. Membership is extended to all who adhere to the aim which constitutes the purpose of the group. Such groups tend to be organized around a single purpose, and group members avidly support that aim and typically exhibit considerable animosity towards people not sharing it. Those outside the group, holding other beliefs, are usually treated collectively as the "opposition" and viewed with general distrust. Commitment to the aim of the group, and maintenance of the strength and purity of that commitment are important activities. Leadership is viewed as a property in fundamental opposition to egalitarianism, and leadership is therefore primarily exercised through exemplary adherence to the group aim and denunciation of the opposition. The sectist cosmology opposes manipulation of individuals, except in terms of the strict defense of the group and its aim. Its ultimate goal is the protection and betterment of the vital property embodied in the aim of the group. The strategy associated with the sectist cosmology is called the collectivist survival strategy.

The region of negative-grid, negative-group in the lower left quadrant is associated with a stable cosmology called "entrepreneural". Here, the experience of the person may be called egalitarian and individual. Each person is viewed according to this cosmology as responsible for his own success or failure, and all are judged according to the same criteria. Group membership is not important, perhaps even a hinderence because it interferes with individual opportunity and responsibility. The function of leadership is to coordinate individual opportunities, and is based on mutual advantage. Action is often justified with the rationale: "If I don't do it, someone else will." It is also generally felt that given enough time and ingenuity virtually any goal can be achieved. Success is often viewed by this cosmology in terms of the ability to exercise authority over others. The preferred strategy associated with this cosmology is called the manipulative individualist strategy.

In the upper left quadrant is the final stable cosmology, labelled "alienation". (With this cosmology Thompson associates the saying, "life is like a lottery", which describes a good deal of its character.) The personal experience of the individual with this cosmology is that he is subject to the dictates of hierarchy, unable to cope effectively both because he is near the bottom of the hierarchical "pecking order" and because he is not granted membership in a support group that could stand in for him. Alone, he is unable to do much other than accept whatever comes. This is the cosmology of life's losers, those who are down and out with little hope for improvement. Things happen over which they have no control, and there is little that can be done once they have occurred. At best one survives, one gets through each day as it comes. The strategy associated with this cosmology is called the individualist survival strategy.

Figure 7.2 presents five characterstics associated with each of these cosmologies in tabular form. These characteristics describe both the way individuals sharing the respective cosmologies are likely to interpret information as well as certain aspect of actions or strategies they will likely expect to be successful. The characteristics are: 1) perception of short and long term risks, 2) typical response to risk in general, 3) concern about the maintenance of social distinctions such as status and association (especially group membership), 4) attitudes towards individual risk taking with the aim of personal reward, and 5) attitudes about the management of needs and resources.

The characteristics presented in Figure 7.2 are, for the most part, self-explanatory. However, the concept of "social risk" in the third characteristic should be explained. Social risk may be thought of as the the likelihood than the social order is in some way damaged. Research has already been referred to which suggests that one of the factors differentiating Austrians most in favor of nuclear power from those most opposed

Alienation

(Individualist Survival Strategy)

1. Only short term risks are perceived with any clarity.
2. No options concerning acceptance or rejection. Risks are a fact of life.
3. No social boundaries to maintain, so little concern for violations. Few, if any, "social risks".
4. Personal rewards are seen as resulting from luck rather than individual risk taking.
5. Individuals are viewed as being able to manage neither needs nor resources. The best, and only, strategy, is survival.

Caste-ist

(Manipulative Collectivist Strategy)

1. Short and long term risks perceived. Long term seen as different but controllable.
2. Risk averse: if risks cannot be completely avoided, they are spread.
3. Many groups and levels of status are recognized, requiring elaborate definitions of membership and policing of violations.
4. Personal risk for individual gain is penalized. Personal risk for collective good is rewarded.
5. Resources but not needs can be collectively managed. The best strategy is to maximize resource use.

Reclusive

(Strategy of Autonomy)

1. Only short term risks perceived with any clarity.
2. If the avoidance of risks requires social obligations, then acceptance is viewed as the lesser of two evils.
3. Because of rejection of both groups and status, there are no boundaries or distinctions to maintain. Few, if any, "social risks".
4. Personal risk taking for individual reward is approved as long as no coercion is involved.
5. Both resources and needs, as well as the conjunction of the two, can be managed. The best strategy is to maximize the conjunction of the two.

Entrepreneurial

(Manipulative Individualist Strategy)

1. Long and short term risks are perceived. Dominance of short over long term maintains expansive optimism.
2. Risks are viewed as opportunities.
3. There is little concern for violation of boundaries (entrepreneurs profit from removal of social boundaries).
4. Personal risk for individual reward is approved, even if involving coercion. Personal risk for collective reward is less popular.
5. Resources and needs can be managed, but not the conjunction of the two. Best strategy is to maximize resources and needs.

Sectist

(Collectivist Survival Strategy)

1. Short and long term risks perceived. Concern for maintenance of the aim of the group causes long term to dominate over the short.
2. Strong aversion to all risks except those involved in the maintenance of the aim upon which the group is mormed.
3. Strong concern about maintaining distinctions between the group and "outsiders" leads to perception of extensive "social risk".
4. Zero-sum game mentality penalizes personal risk taking for individual gain. Personal risk taking for collective gain is enthusiastically rewarded.
5. Needs but not resources can be managed. Best strategy is to maximize the conjunction of the two by minimizing needs.

to it was the concern about the social and political implications of increased reliance on the technology. (Otway, 1980, p. 42)

It should come as no surprise that those cosmologies which are based on a history of personal experience in which clearly defined groups or hierarchical patterns of status differentiation have proven important are more likely to perceive dangers to the social order. An individual with a caste-ist cosmology, for example, will be aware of a number of different groups and social actors involved in the management of the technology in question and the response to the hazards it presents. He will expect different response from experts and laymen, from public officials and industry representatives, from citizen activists and politicians. He is likely to be quite concerned that conflicts about hazardous technologies may so interfere with "normal" decision processes that society may become "ungovernable". An individual sharing the sectist cosmology, on the other hand, might respond to the same technological hazard in quite another way. Such a person might be convinced that the extent and pervasiveness of technological hazards in general present a constant and frightening danger. He would likley lump government and industry together as the perpetrators of the problem, and feel that anyone not advocating widespread and immediate reduction of such hazards to be on the "other side". ("If you are not part of the solution, you are part of the problem.") Such people might feel that there is a present and continuing danger to society in allowing the "conspiracy" of government and industry to continue to despoil the world, and that this conspiracy should be opposed by whatever means possible. Both caste-ist and sectist oriented individuals perceive social risks in the issues raised by technological hazards. But the social risks they perceive may be quite different.

The classification of people according to these cosmologies and the associated strategies of action can contribute towards anticipating how they are likely to respond to a particular technological hazard.

Summary and Conclusions

This chapter presents an assessment of what are referred to as systemic influences on modern technological conflicts. The aim is to develop a consistent picture of the cultural context within which technological hazards should be seen. But rather than attempting a broad examination of society in general, several key developments in modern industrial society are highlighted and their consequences for the management of technological hazards examined.

These may be summarized in the following manner to suggest how they interact with ideas presented in other chapters.

> There are good reasons to carefully examine human activity and its consequences whether one advocates rapid technological advance (technocentrism) or protection of the environment (ecocentrism).
>
> But analysis of the consequences of the application of technology always leaves unanswered questions and the possibility of alternative explanations. (See the discussion of the unavoidability of professional judgment in formal analyses of technological hazards in Chapter 3.)
>
> This permits the situation in which experts argue different positions in technical conflicts. Furthermore, scientific evidence is increasingly cited by politicians to support programs of action. Thus, it is common to find scientific and technical analyses as a part of the debate about technological issues, and sometimes apparently even the source of technical conflict.

> The patterns of decision making concerning the development and application of particularly hazardous technologies often corresponds to the dominant pattern of social domination in modern industrial societies. Conflicts about technological hazards can therefore sometimes be viewed by some people as just another battle in a larger war against domination.
>
> Personal experience is likely to lead different individuals to perceive the same situation differently. Although this assertion sounds trite, it takes on a rather more fundamental meaning with respect to cultural patterns like organizational interaction or manipulation and submission. Thus, the individual's cosmology, derived from his experience of cultural patterns, is likely to influence both his interpretation of information (related to the discussion of cognitive attributes of technological hazards in Chapter 5) and the strategies preferred (including the interdependence of problems and opportunities, and the advisability of cooperative response as discussed in Chapter 6).

Thus it may be seen that even though the technical attributes determine the parameters of physical cause and effect, and individual reaction and collective response both strongly influence public opinion and the structure of intervention programs, in some situations the systemic influences described in this chapter can dominate the overall response to technological hazards. An attempt to be rational in managing technological hazards cannot omit the cultural influences on the way technologies and their hazards are viewed and evaluated.

8 ORIENTATIONS AND OUTLOOKS REFERENCES

There can be no doubt that some technologies are associated with hazards. Furthermore, the problems created by these hazards are often particularly vexing. In the simplest cases they can be dealt with by modificating the machinery, equipment or materials used so as to eliminate or reduce the exposure to the hazard. In such cases, it is probably enough to decide what to do on the basis of purely technical considerations, relying on the professional judgment of the engineer. But there are other situations, generally when the consequences of the hazard are borne by people not occupationally associated with the activity producing the hazard, where social and political concerns must be given equal weight with the technical attributes in the determination of how to respond. Sometimes the social and political concerns of people not otherwise involved prove to be most important element, completely overriding other factors. For example, the compensation of injuries to third parties attributable to business activity has been imposed on enterprises in industrialized countries, regardless of the financial position of the individual enterprise. This means that marginal firms, those barely meeting their costs prior to the necessity of compensation, could be forced out of business, especially in a large accident involving many people. The social concern about the livelihood of those injured is clearly given priority over the technical considerations, including the costs of operation or even survival of the firm.

But what makes the management of technological hazards per se vexing is the necessity of integrating not only the technical and economic factors but also relevant social and political concerns in a single decision about how to respond to a given hazard. Both the technical factors--the physical equipment and materials, and their organization into a production system--and the economic factors governing the operation of that system can be characterized by rather well defined models. That is, the elements involved can be more or less precisely defined, and the interactions among those elements specified with a fair degree of clarity. Not so for the social and political factors; they are difficult to define, and the interactions among them and with the technical and economic factors are poorly understood. This is the principal source of vexation. Although there are unavoidable uncertainties confronting the analysis of the technical and economic factors associated with technological hazards, the analytical approaches for dealing with them are relatively clear. Probabilty theory and statistical analysis can be applied to obtain estimates of likelihood, since the uncertainties deal with alternative outcomes of well-defined systems and the constraints of these theoretical approaches can be reasonably well met. But the uncertainties associated with the social and political concerns are of a much different nature: it is often unclear which elements play a role, much less the ways they interact with one another or with the technical elements. There are often a variety of different kinds of concerns being considered simultaneously in different decision arenas. Examples abound--nuclear power, saccharin, laetril--where it is difficult to know whether scientific and medical evidence, custom of usage, or symbolic meaning will turn out to be more important in the ultimate decision.

Hazards and how to respond are often contemplated at several levels of aggregation or focus, with different elements receiving emphasis at each level. Sometimes, the focus of one level leads to a particular way of thinking about the situation that does not translate well to other levels (for example, it is often difficult for the person designing a piece of equipment, who must necessarily focus on the technical artifacts of the system, to incorporate consideration of the possible impact of a new design on social conditions through alteration of work patterns). The well defined and precise analysis applicable to the technical and economic factors is often inapporpriate for the examination of such social and political factors because the uncertainties concern primarily the identification of important elements and their interaction as much as the measurement of their respective influence. Of course, there are scientifically sound approaches to such uncertainties as well, but they necessarily involve fundamentally different kinds of input and employ different methods of analysis.

Further, there are likely a number of different interests relevant to any decision of importance. Each party at interest will try to insure that its view is considered, or even that it dominates, if possible. Thus, one can find a kind of conceptual incompatability between the technical and the social and political factors on the one hand, and a competition among interests on the other. Yet, decisions about response to technological hazards must be made, and must often refer to social and political factors as well as technical and economic data. One result is that such decisions are often made in an atmosphere of confusion and controversy, because of the necessity to balance factors for which the available analytical concepts are incompatible or competative.

Basic Concepts and Approaches in the Analysis of Technological Hazards

Over the past several decades a series of analytical methods have been applied to the task of controlling technological hazards. Some of these methods are used by government agencies as they fulfill their regulatory responsibilities. Some of them are used by business enterprises as they attempt to avoid or reduce losses in the pursuit of profits. These various concepts and approaches have served as the starting point for this study. Because of their variety, as well as the diversity of those concepts subsequently singled out for application in terms of the conceptual framework developed in Chapter 4, it seems advisable to present an overview and summary in the concluding chapter.

Unfortunately, the word "risk" is used in rather different ways in different contexts. In general economics, risk usually refers to the possibility of incurring profit or loss. But in "benefit-risk analysis", one of the methods of risk management, the implication is that risk is something to be balanced against benefits, and therefore somewhat similar to potential losses. This usage corresponds to that in many firms, where "risk and insurance management" refers to the reduction of the potential losses and is usually considered to be a part of financial management. One result of such terminological imprecision is confusion about exactly what is meant when referring to risk management. There may be agreement about the general aim, but there is often difference of opinion about the application of specific procedures.

Because of this confusion, the study therefore specifies a

set of definitions for the following key terms: hazard, exposure, risk, effect, reaction, and response. In brief, risk combines the possibility of exposure to some hazard from one or more sources with the likelihood that this exposure leads to the occurrence of some undesirable effect. Reaction refers to the intellectual and emotional state stimulated by knowledge of this situation, and response refers to the behavior undertaken as a result. These key concepts provide the basis for the subsequent description of the management of technological hazards and the human and organizational interactions at various levels of social aggregation.

A basic idea underlying risk analysis is that technological hazards should be seen in terms of: 1) the nature and extent of the source of danger or potential damage; 2) the chain of events, pathways and processes that connect the cause(s) to the effect(s); 3) the relationship between the characteristics of exposure and of effect; and 4) the nature and strength of individual and group response to the sequence from cause to effect. Information about each of these four different topics is required for the intentional and rational management of technological hazards.

The methods commonly used to describe technical attributes of technological hazards may be illustrated by safety and reliability engineering. Safety engineering concepts which are important for the overall interpretation of data in the management of technological hazards include the following (McCormick, 1981):

> Schematic system diagrams (usually showing the functioning of the system in terms of the flow of a signal or material from system input to system output),

Minimal cut sets (the minimal events necessary to interrupt the flow from input to output),

Common fault failures (failure of several elements due to a single event),

Common mode failures (failure of identical, redundant components due a single event),

Fault tree analysis (the diagramming of the possible causes of a particular component failure through detailed analysis of the events contributing to the failure under examination), and

Event tree analysis (the diagramming of the possible consequences of a particular event through detailed analysis of the system elements).

The quantitative estimates of system failure which these and other concepts have made possible constitute a considerable improvement in the analysis of technological hazards. Nevertheless, the application of systems safety engineering is not without controversy. For example, the U.S. Reactor Safety Study (U.S. Nuclear Regulatory Commission, 1975), has been the subject of extensive criticism despite the extensive effort that went into the study. Among the objections to the report were: the system is simply too complex to quantify, the data base for component failure rates does not support the calculations, improper and incorrect statistical procedures were applied, common cause failures were inadequately treated, low probability events are intrinsically impossible to quantify, and human behavior was inadequately treated. (Lewis, 1980) But despite such criticisms, safety engineering methods are generally felt to provide a much better basis for decision making than would be available without their application.

The quantification and precision offered by these methods make it possible to identify (and therefore also to criticise)

the principal assumptions in major analytical efforts. Although safety engineering brings rigor and precision to the description of the technical attributes of technological hazards, the analyst must still rely on professional judgment both to draw the bounds of the system being examined and to determine the level of detail to be included in the analysis. The quality of the analysis will be evaluated according not only according to the accuracy of his calculations but also by the degree to which his judgment corresponds to professional consensus in relevant fields of expertise.

The second major analytical basis for the management of technological hazards is cost-benefit analysis. (Mishan, 1976; Sassone & Schaffer, 1978; Hofmann, 1981) The approach is based on the simple idea that the advantages and disadvantages of a contemplated action should be compared before it is undertaken. Cost-benefit analysis accomplishes this comparison through the use of a numeraire, usually a monetary valuation of the costs and benefits associated with the action. But the application of the cost-benefit approach generally goes beyond this simple notion to evaluate the costs and benefits to all members of society as well as the individual or group that is to initiate the action. Thus, cost-benefit analysis relies on concepts like opportunity cost, social benefit and excess social benefit over cost as well as on cost, revenue and profit.

In the ideal model, all effects or consequences can be bought or sold on the market and their prices reflect full costs, no group or individual seller is able to exert disproportionate influence, and transaction costs are insignificant. However, because these conditions are never fully met in reality, the analyst must deal with a series of approximations, each requiring the exercise of judgment. The two most important areas of judgment in cost-benefit analysis are probably the selection of the consequences to be included and their valuation, and the determination of the discount rate to be used in assigning value to future events and conditions. Because of the importance of professional judgment in cost-benefit analysis, it is essential that the assumptions and procedures used in an analysis be made available for criticism and review by others than those conducting the analysis.

The amount of intellectual effort that has been devoted to refinement of the cost-benefit approach is probably on a par with that devoted to systems engineering approaches. Similarly, there are those who advocate its wide application to all manner of public and private decisions and others who damn its inadequacies as wrongheaded and a mistaken basis for proper decisions. For example, some analysts attempt to show the limits to applications of the approach (Pearce, 1976, explores situations in which the "optimal" level of output of certain environmental pollutants would be below those dictated by initial cost-benefit criteria). Others describe the need to develop procedures promoting greater accountability and public participation in procedures which use cost-benefit approaches as a basis for regulatory decisions. (Barum, 1980) But despite the difficulties confronting the application of the methodology, the idea of comparing the advantages and the disadvantages of contemplated action is so fundamental that some form of cost-benefit analysis will certainly remain a key element in the management of technological hazards.

The third analytical basis for the management of technological hazards deals with the analysis of alternative action programs, usually from the perspective of society as a whole. (Battelle, 1976; Porter et.al., 1980) The general approach is refered to in this study as technology assessment/environmental impact analysis (TA/EIA), and includes four components: analysis, forecasting, evaluation and recommendation. The approach can be applied to broad questions and generic applications of technology (e.g., technology assessment) as well as to site-specific applications of a specific technology (e.g., environmental impact assessment). Such analyses are generally performed before the technology in question is implemented, with the intention of anticipating undesirable consequences so as to avoid them. Considerable emphasis is given to the systematic examination of the effects of the technology on the social system and to to consequences that are unintended, indirect or delayed. Thus, the analysis refers not only the technology, its associated hazards and the chains of cause and effect leading to undesired consequences, but also the characteristics of the population exposed, its likely response and the overall impact on the society.

The concept of TA/EIA essentially incorporates many of the methods of both systems engineering and cost-benefit analysis in a framework which also examines the likely responses of relevant groups in society. Thus, to the ideas of the other approaches are added the concepts like social description, impact analysis and evaluation, policy analysis and communication of results. Participation by lay citizens in certain of the steps is often given considerable emphasis in TA/EIA approaches. Thus, to the difficulties confronting the application of the more technically oriented analyses are added those of identifying and working with relevant groups in society. It is easy to see how such an effort can become embroiled in the political maneuvering associated with proposed or existing action programs.

Some of the difficulties of incorporating all these different factors are less in the fourth major analytical basis, the application of what is commonly called risk and insurance management within the individual firm. (Green & Serbein, 1978; Carter, 1982; Olson & Simkiss, 1982) A self-interested firm examines the possibility of losses due to technological hazards, and compares the costs associated with prevention programs with various forms of financing the potential loss. Of course, this usually entails assigning somewhat different importance or meaning to various elements in the chain of events from source to consequence than is the case for public assessments, but the managers of a firm do perform judgments that are in principle quite similar to those in societal analyses. The principal differences, however--the restriction of attention to the potential losses confronting the firm and the actions that might minimize the effect of these on the firm--often lead to quite different conclusions.

The focus on minimizing losses brings one attribute of technological hazard into clear relief which often remains clouded and vague in many societal analyses. Some potential loss-producing events are associated with activities intended to produce income, others are not. In risk and insurance management, as practiced by many enterprises, the former are called speculative risks and the latter pure risks; a distinction clearly derived from the emphasis on profit and loss. Societal analyses often attempt to deal only with negative consequences and ignore the benefits. When they are included, it is often unclear exactly how the two should be balanced against one another. Such problems are usually much less pronounced in analyses performed by private enterprises for their own purposes.

Of relatively greater importance for a firm's risk and insurance manager is the issue of how to finance potential losses. In general terms, a balance must be chosen among four basic options: loss prevention programs, transfer of risk to some other entity by contract, assumption of risk and payment of losses either out of cash flow or an internal fund drawn from working capital or retained earnings (self insurance), and purchase of insurance from a commercial company. The determination of the likely impact of these various options on the cash flow of the firm, its market position or the profit and loss statement is not a simple matter.

In many respects the tasks of risk and insurance management are the same as those confronting analysis focussing on the society as a whole. Although the criteria of measurement and the ultimate goal may be simplified by the restriction of the analysis to the consequences for the individual firm, the technical uncertainties and the difficulties of applying the various methods remain. But since the source of many technologial hazards may be found in large corporations, it is important to understand not only how they are analyzed from the perspective of society, but also how such hazards are viewed within these firms.

A Conceptual Framework for the Management of Technological Hazards

This study shows that the literature of risk analysis reflects a discussion about technological hazards and response to them may be said to have gone through three phases. The first emphasized mostly technical aspects of hazards and response to them. The second devoted more attention to the evaluation of exposure to these hazards, especially in comparison to other hazards. Most recently, greater emphasis has been given to the conflicts associated with such hazards, including programs of response to them. Thus, one can say that there has been a trend from the technical to the political. This study presents a conceptual framework which is consonant with that development. This framework incorporates the existing approaches, outlined above, as well as providing a simple model which also refers to the necessary social and political judgments.

Typically, one finds four sets of individuals, groups and organizations that are relevant for the management of a technological hazard: the technocrats (who operate the technology), the public officials (who influence the behavior of the technocrats and thereby attempt to exercise control over the technology), the politically active (who attempt to establish laws and administrative regulations controlling activities they consider important) and the otherwise uninvolved individuals (who often bear the brunt of the exposure to the hazard).

The simple identification of classes or sets of individuals, groups and organizations is not, of course, an adequate basis for the management of technological hazards. Even the additional observation that each usually exerts influence of a particular kind is insufficient. One also needs a description of the way individuals and groups or organizations reach decisions in such a pluralistic setting. The study relies on a model in which the individual focusses initially on technical matters and then shifts to the potential interactions at various levels of social aggregation. The individual begins by marshalling facts and arguments, then attempts to secure support from his or her subordinates, colleagues (or peers, when functioning in a group rather than a formal organization), the constituency of the organization or group (those with whom information, goods or services are exchanged), and finally from the broader public. This sequence is combined with the typical ways that support is sought (i.e., through exercise of authority, through exchange, or through persuasion) to suggest possible ways of securing support for a program of action. This model provides a basic characterization of the decision process which is based on individual volition, but which includes interaction with individuals, groups and organizations at various levels of social aggregation.

The final component is a framework which sets out the principal topics relevant for the management of technological hazards. It is based on four main tasks: specification of causal relations, characterization of priorities, assessment of capacity to respond, and design and implementation of intervention measures. The framework is conceived so as to place not only the information concerning relevant technical and scientific judgments in proper perspective, but also the information required for political judgments needed to evaluate the consequences and to implement intervention measures.

The specification of causal relations is characterized as

being made up of the identification of hazards and consequences, and the estimation of the likelihood of occurrence. The characterization of priorities is seen to consist of the assessment of awareness and conflict and the evaluation of the consequences. The assessment of response capacity, in turn, is described as the analysis of existing organizational capacity and potential professional capability. The design and implementation of intervention measures is described in terms of the specification of the scope of the program, the selection of the structure for management and control, and implementation.

Taken together, these three sets of ideas--typical patterns of interaction, decision making in a pluralistic setting, and the conceptual framework--constitute a basis for managing technological hazards. They are offered in the spirit of an analytical framework rather than a theory. That is, the intention is to develop a set of ideas with which a rational and intentional response to technological hazards can be developed. Whenever possible, concepts are based on actual experience with technological hazards. But the intention was not to test these concepts against actual decisions. The aim of this study is to suggest how information for the decision about how to respond to technological hazards ought to be organized, not how such decisions are made today.

Social and Political Factors Influencing the Management of Technological Hazards

In the final chapters, factors particularly relevant to the political judgements involved in managing technological hazards are examined in greater detail, so that the conceptual framework might incorporate such influences with greater sophistication. The influences on political judgment are examined on three levels: individual reaction to technologies and the hazards associated with them, collective response to technological issues, and systemic attributes of modern technological hazards.

The level of individual reaction is in some respects the easiest to incorporate, simply because more research has been done about technological hazards and reactions to them. Thus, the study can review research results relevant to individual reaction to technologies and their hazards, whereas for the other two levels the sources of ideas are more disperse and sometimes previously not applied to the issues associated with the management of technological hazards.

At the level of individual reaction, risks are conceptually universal, unmeasurable and uncontrollable. (Ravetz, 1980) An individual tends to treat each personal reaction to a technological hazard, like all other actions, as another element in his total experience. Thus, the analysis of individual reaction to technological hazards is in some respects like putting all human experience into a single schema. But not only are hazards universal for the individual, they are incalcuable. Major hazards, where damages would be great but the likelihood of occurrence is probably quite small, are often treated by people as if they were a gigantic lottery. The outcome depends on luck rather than a calculated gamble. Even for more "ordinary" events, like an automobile trip, the information available is generally inadequate for a proper calculation of the odds. Statistics about automobile travel may be known, but probably not for the precise route to be taken, time of day, etc. Finally, hazards are conceptually uncontrollable. It can never be know whether enough has been done to prevent a hazard from occurring. Even after an event has occurred, it is impossible to be certain how much more would have had to be done to prevent it, or whether that level of activity would have been "reasonable". Nor is this difficulty simply a matter restricted to the unsophisticated reactions of laymen. Even after the loss-of-coolant accident in the Unit Two Reactor at Three Mile Island, it is impossible to tell whether the estimates for such accidents in the U.S. Reactor Safety Study are accurate or not.

In other words, an individual is necessarily confronted with uncertainties when thinking about technological hazards. However, there are a number of typical ways of coping with such uncertainty. In particular, three heuristics, or simplifications, are commonly used to enable the individual to draw conclusions based on incomplete or uncertain information. (Tversky & Kahneman, 1974) The first, representativeness, is used to infer properties based on apparent similarity. The second, availability, bases conclusions on whatever attributes can be brought to mind. In other words, it relies on familiarity. The last, anchoring and adjustment, establishes a preliminary estimate and then modifies that value to better approximate the individual's true feeling.

Research into judgments under uncertainty led to the recognition that the assumptions about individual preferences in the most common model of individual decision making, expected utility theory, do not correspond to the way people actually perform such judgments. An alternative, prospect theory, has been developed to incorporate some of these differences. (Kahneman & Tversky, 1979) It differs from expected utility theory (most easily characterized by the behavior of "economic man") in two important ways. The value of a future outcome is not viewed simply as the intrinsic value of that outcome multiplied by its likelihood of occurrence, but is rather seen as a positive or negative deviation from some neutral outcome. The way value is assigned to this deviation has two important properties. The first property is that the value, or importance, assigned to this deviation tends to be concave for positive and convex for negative deviations. This means negative deviations of the same "magnitude" are assigned greater importance than positive deviations. The second property is that events with very small likelihood of occurrence are given greater relative weight compared to events that are quite likely. These assumptions in prospect theory lead to quite different response to catastrophic events (which by definitiion have very large consequences but generally have very low likelihood) than would be predicted by expected utility theory.

This is an important point. It is altogether possible to come to different valuations of potential outcomes than those derived from standard economic analysis, yet not be arbitrary, inconsistent or illogical. Perhaps more important is the fact the positions taken by those opposing technological developments are often better explained by prospect theory than by expected utility theory. This can be true for general reaction to technological hazards as well. (See: Slovic, Fischhoff & Lichtenstein, 1980)

A second line of research has examined reaction to technologies and their hazards in terms of attitudes instead of judgments. (Otway & Fishbein, 1977; Otway, 1980) Attitudes are typically based on a number of beliefs about the object, person, concept or symbol in question. Personal sensory perception is the strongest influence over such a belief system, so that what a person can observe himself is a powerful influence over his attitudes. According to the model, the attitudes a person has determine his intentions with respect to the object, and his intentions aggregate into his behavior towards it. It is important to understand that when people respond, say in public opinion polls, about their reactions to technological hazards, they are generally expressing their attitudes in response to the questions put to them. Thus, one must be careful to distinguish among beliefs, attitudes, intentions and the actual behavior towards the object of the question.

This research suggests several conclusions that are useful for rational and intentional treatment of technological hazards. First, because technical information can seldom be confirmed through personal experience, such information is often less convincing than information an individual is able to observe for himself. Such information can be expected to overcome previously held attitudes only when there is no disagreement concerning its validity and veracity. Even then, technical information may prove quite slow in changing attitudes. Further, because attitudes are determined by an aggregation of beliefs about the object in question, technical information may not address those factors upon which the attitudes about the object are based.

Thus, in examining psychological and emotional reaction to technological hazards, one not only needs to consider whether the individual is performing the judgments involved based on assumptions other than those associated with the standard economic premises of expected utility theory, but also which attributes of the technology and its hazard influence the attitudes held by the individual. If there is disagreement about the acceptability of a technology or the exposure to a hazard, one will likely find differences on one or both of these points. Furthermore, another line of research there may be differences in the basic values upon which broadly defined preferences are based.

The postulated emergence of a "post-materialist" orientation (Inglegart, 1971), with its greater emphasis on the quality of life and "self-actualization", could also be a source of disagreement and conflict with respect to technological hazards. The post-materialist orientation is consistent with a rejection of the idea that technological progress per se is a good idea, and with a much more critical position vis a vis technologies in general. There can be little doubt that such ideas are given much more credance than they received a few decades ago. However, the manner in which the formal model is formulated casts some question on the reliability of conclusions about these postulated value changes. Thus, whether or not the apparent rejection of technological progress by some people is a result of a widespread shift in values is difficult to say. But in any case, these issues are important for the management of technological hazards, and are also related to the systemic attributes of modern technological hazards.

The discussion of collective response to technological issues in this study is in a somewhat weaker position than that of individual reaction to technologies and their hazards. There has been considerable research conducted which reveals important influences on individual reaction to technologies and their hazards, but there has been much less research which addresses influences over the response of groups and organizations as such. Of course, there have been innumerable case studies of technological conflicts, and there is a large economic literature on the behavior of groups (as well as, of course, the treatment of various collectivities, such as firms, as economic actors). But the available research focussing on collective behavior in technological conflicts is not particularly well suited to the approach of this study. That is to say, it does not fit neatly the separation of behavior into three levels: individual reaction, collective response, and systemic factors. Thus, the treatment of collective response in a way enabling the further development of the conceptual framework is based on concepts drawn from research not directly addressing technological conflicts.

The treatment of collective response to technological issues in the study follows roughly a stylized model of decision making. It begins with the specification of the parameters limiting the choice, identifies the nature and degree of interdependence between or among relevant actors, examines structural factors influencing the strategies available to those actors, and finally, explores ways of coordinating the action of different parties.

The parameters of choice with respect to technological conflicts are seen as the product of the intersection of three sets of competing influences. The first set is the tension between science, informed by consensual knowledge of cause and effect relations, and politics, informed by normative purpose, negotiated priorities and available capacity. The second set of influences is the tension between the need to respond collectively to shared problems and opportunities, and the desire to maintain autonomy and flexibility of action. (Ruggie, 1975, described these first two parameters.) The third set of influences on the parameters of choice is the tension between the need to stimulate the development of new technologies and the necessity of coping with the consequences of using them. Each response to technological issues may be seen to reflect some position with respect to each of these parameters of choice.

The key concept in the treatment of collective response is the idea of interdependence. (The concept is introduced, in slightly different form, in Ruggie, 1975, pp. 561-564.) It is described in terms of the type of collective situation confronted by the relevant actors (e.g., cognitive, opportunity cost, constraint, or deprivation) and the "location" or functional distance of the basis for collective action from the central concerns of the group or organization (e.g., external, external manifestation, "commons", or internal). Each collective situation represents an agreement by some set of actors that one particular configuration describes their position. This agreement may change as knowledge of the relevant cause and effect relations increases, as configurations of interdependence expand or contract to encompass different sets of actors, or as the objectives and capabilities of relevant groups or organizations change. Cooperative action is therefore not a matter to be decided, but to be cultivated and managed.

The selection of strategy by a group or organization is described in terms of reaction to the "causal texture" of the environment within which the actor is located. (Emery & Trist, 1965) Four types of causal texture are identified (placid randomized, placid clustered, disturbed-reactive, and turbulent field) together with the corresponding optimal strategies (respectively: trial and error tactics, strategic goal setting, operational action and response, and matrix organization). Each type of causal texture not only represents an increase in the complexity of the analysis and planning required by the actor, but also corresponds to an increase in uncertainty concerning the information necessary to do so. This description of the structural influences on strategy choices by groups and organizations underscores the importance of interdependence in determining collective response to technological issues.

Four major factors are considered important with respect to coordination of activity: the instruments of coordination (common framework, joint facility, common policy, centralized direction, hierarchical agency), the role of key individuals (e.g., representatives of government, representatives of private associations, executive heads of organizations, high bureaucratic officials, employees of the mass media), the exercise of influence (e.g., iniation, veto, control, brokering) and the functional orientation of the social unit (informational, managerial, executive, and special interest). (Ruggie, 1975; Cox & Jacobson, 1973)

The treatment of collective response to technological issues in this study suggests why collective action is as dependent on the respective aims and interests of the various actors as on their common problems and opportunities. Thus, one may expect that unanimity in response to technological issues would be the exception rather than the rule. It also suggests why purely technical analyses of technological hazards which omit the "political" motivations of relevant groups and organizations are often treated as somewhat irrelevant by decision makers as well as lay persons.

The discussion of systemic attributes of modern technological is similar to that of collective response to technological issues in that it draws upon concepts from diverse sources. Rather than attempting a broad description of society in general, several key attributes of modern industrial society are highlighted and their consequences for the management of technological hazards described. Although these attributes may at first appear haphazard and arbitrary, they constitute a consistent influence in modern industrial societies.

In general, it may be said that there are two ideological bases for the critical review of the use of technology: ecocentrism and technocentrism. (O'Riordan, 1981) Both the ecocentric and technocentric modes of thought call for the careful examination of human activities. But, the reasons for doing so differ according to the two positions: ecocentrism advocates the minimization of the effects of human activity on nature, technocentrism aims at the optimal use of nature. But both ideological extremes vis a vis technological development encourage the critical examination of technological activities. This constitutes an important and prevalent predisposition with respect to technologies and their hazards.

Another attribute of modern industrial society which is important for the management of technological hazards is the politicization of science and technology. The necessity of public financial and other support (Bermant, 1982), the attempt to institute legal and administrative control over adverse consequences of the conduct of research and the use of technology (Nelkin, 1979), and the emergence of scientific and technical expertise as central elements in public conflicts about technological issues (Nowotny, 1979) each contribute to the view that science and technology, like other elements of society, are subject to social and political controls. But even more important, they lead to the feeling that science and technology themselves are controversial.

But technological conflicts can be seen in some instances to be merely another manifestation of a much wider and more fundamental set of social conflicts described by Touraine (1971). Technical conflicts often follow the typical pattern of social dominance in modern industrial societies: the individual is forced to participate in certain activities and social structures regardless of personal desires. Exposure to technological hazard generally follows this pattern of dependent participation. The individual does not choose to be exposed; the hazards are imposed without consulation. In fact, the people actually bearing the consequences of such hazards may not even know exactly who is involved in the decision or where it takes place. But as has always been true for systemic social conflicts, there is a typical response in opposition to the phenomenon: the demand for widespread participation in all decisions. Thus, the behavioral response to a given technological hazard may be as much determined by the attempt to deal with the broader system of control and decision making as by the particulars of that hazard itself.

Finally, shared patterns of personal experience can

influence the way the individual perceives problems and opportunities. The degree to which one is subject to the dictates of personal pacts with others and to restrictions deriving from membership in groups or organizations, together with the degree to which one can use such pacts and membership to impose one's wishes on others as opposed to submit to their will, are taken as important determinants of both the way one sees the world and the strategies one prefers. (Thompson, 1980a) Five patterns of experience, or cosmologies are identified (alienation, caste-ist, sectist, entrepreneurial and reclusive) together with the strategies associated with them: (respectively: individualist survival, manipulative collectivits, collectivist survival, manipulative individualist, and autonomy).

One can conclude that careful analysis of human activity seems to be called for no matter which ideological extreme one chooses with respect to technological development. But there are typically competing interpretations of the relevant scientific and technical information. In many cases, such alternative interpretations have received considerable attention in public debates about technical conflicts. This combines with the increasing dependency of research and technological progress in general on political decision makers to give science and technology a decidedly political appearance. Further, the patterns of decision making associated with such conflicts often correspond to the dominant pattern of social domination in modern industrial societies, so that some groups and individuals respond to technological issues in terms of broader social conflicts rather than the specifics of the hazard involved. Finally, the overall experience of different individuals can lead them to perceive the same situation in quite different ways, leading to decidedly different strategies of action. Thus, the possibility that systemic attributes dominate the response to technological hazards cannot be dismissed a priori. A rational approach to managing technological hazards, and in turn its contribution to a preventative environmental policy, must acknowledge them and consider their influence.

These ideas by no means exhaust the possibile societal influences on the management of technological hazards. But they do provide a description of systemic influences that compliments those of individual reaction and collective response. Taken together, the treatment of these three levels of influence can provide the basis for further refinement in the application of the conceptual framework.

Prospects for the Management of Technological Hazards

It has already been mentioned that analysis of possible response to technological hazards takes place both in government agencies and business firms, as well as in groups of private citizens. The differences in approach, and the ways similar approaches are differentially applied, will no doubt continue to be a source of conflict in responses to technological hazards.

One important difference between government and business assessments is the overall goal of the analysis. Analysis within a private enterprise usually is aimed at balancing the costs of prevention and reduction with the potential losses associated with hazards, as these accrue to the firm. Analysis taking the public perspective generally examines the balance of gains and losses falling on a variety of different groups, including uninformed individuals and sometimes even future generations. Private analysis usually limits its scope to the physical cause and effect, and to the financial consequences for the firm of preventative action versus response after something has happened. Public analysis generally looks not only at physical cause and effect, but also at individual and group reaction to the hazard and the technology producing it, to wider effects on the social system, and to the problems of implementing a program of intervention. Public analysis generally views financial costs as distributed over the entire population.

This brief characterization of the differences in the overall aims of public and private assessments of technological hazards suggests that the catagories described here would be of greater use in public assessments than in private ones. However, it should be noted that whenever there is the possibility that the political attributes of the hazard in question outweigh the technical side of the situation, then consideration of the full set of factors suggested here would to the advantage of an individual enterprise as well.

It is sometimes suggested that much of the cumbersome structure of regulatory government could be replaced with formal analysis of technological hazards and implementation of intervention measures in the private sector, thereby giving relief to taxpayers. The analysis here, however, suggests that such a substitution would not be feasible. Uncertainty is at the very core of the management of technological hazards; uncertainty about alternative interpretations of data and explanations of relevant chains of cause and effect, uncertainty about the consequences for various groups and how these consequences should be evaluated, uncertainty about the effectiveness of various interventions and the manner in which they should be organized. Each of these points can lead to conflict. Each can be hotly contested by individuals and groups affected by a particular technological hazard, or by those attempting to influence the overall decision structure with which technological progress is influenced. It is most unlikely that a formal analytical procedure could be developed that would satisfy the various parties potentially at conflict. Indeed, one must conclude that the structure of regulatory government is cumbersome exactly because it must balance such competing interests. Although analysis of technological hazards is essential for rational response to them, it is naive to expect a formal analytical procedure to replace a process which assigns an official the responsibility for the final choice among competing interests.

No doubt government can be made more efficient and less cumbersome. But it would be a mistake to try to replace political responsibility with formal analysis. One can attempt, and that is the aim of this study, to increase the rationality of the basis for such decisions. But the clarification of the factors at play and their interactions with one another need not and should not dilute the political responsibility associated with the final decision.

REFERENCES

Ashby, W. Ross. 1960. Design for a Brain. London: Chapman & Hall.

Barum, Michael S. 1980. "Cost-Benefit Analysis: An Indadequate Basis for Health, Safety, and Environmental Regulatory Decisionmaking" Environmental Law Quarterly Vol. 8, No. 3, pp. 473-531.

Battelle. 1976. Technology Assessment: Methoden, internationale Entwicklungen, Auswirkungen auf die unternehmerische und staatliche Planung. Bericht im Rahmen eines Gruppenprojekts. Frankfurt: Battelle Institut e.V.

Bechmann, G.; Frederichs, G. & Gloede F. 1981. Technology Policy and Value Change--On the Acceptance of Energy Systems. Research Proposal prepared for the Workshop "Analysis, Evaluation and Acceptability of Hazardous Technologies", 14-16 December. Berlin: International Institute for Environment and Society.

Bermant, Gordon. 1982. "Justifying Social Science Research in Terms of Social Benefit" in: T. Beauchamp, et.al., (Eds.) Ethical Issues in Social Science Research. Washington, D.C.: Johns Hopkins Press. pp. 125-143.

Biervert, Bernd; Fischer-Winkelmann; Wolf, F. & Rock, Reinhard. (Eds.) 1978. Verbraucherpolitik in der Marktwirtschaft. Reinbek bei Hamburg: Rowohlt Taschenbuch Verlag.

Bowman, James S. 1976. "The Environmental Movement: An Assessment of Ecological Politics" Environmental Affairs Vol.4, No.4, pp. 649-667.

Buchanan, James M. 1965. "An Economic Theory of Clubs" Economica Vol. 32, No. 1 (February) pp. 1-14.

Burton, Ian & Whyte, Anne V. 1980. Environmental Risk Management. SCOPE 15. Chichester, England: John Wiley & Sons.

Burton, Ian; Klinan, Mel; Powell, David; Schmidt, Larry; Timmerman, Peter; Victor, Peter; Whyte, Anne & Wojick, Joanne. 1981. The Mississauga Evacuation; Final Report to the Ontario Ministry of the Solicitor Ceneral. Toronto: Institute for Environmental Studies, University of Totonto.

Business Week. 1979. "Interventionist Government Came to Stay" (3 September) pp. 39-43.

Business Week. 1982. "Manville may drive Congress to action." (13 September) pp. 22-23.

Business Week. 1982. "Creditors and Investors Wait Out the Confusion" (13 September) p. 23.

Carter, Robert L. 1982. "Risk Management Under Changing Economic Conditions" The Geneva Papers on Risk and Insurance Vol. 7, No. 22 (January) pp. 3-12.

Coleman, James S. 1982. Free Riders and Zealots. (Mimeographed) Berlin: Wissenschaftskolleg zu Berlin.

Commission of the European Communities. 1979. The European Public's Attitudes to Scientific and Technical Development. Brussels: The Commission. (XII/201/79-EN)

Coppock, Rob. 1977. "Decision Making When Public Opinion Matters" Policy Sciences Vol. 8, No. 2, pp. 135-146.

Coppock, Rob. 1980. "A Synopsis of Discussion and Suggestions for Further Action" in: Dierkes, Meinolf; Edwards, Sam & Coppock, Rob. (Eds.) Technological Risk: Its Perception and Handling in the European Community . Cambridge, Massachusetts: Oelgeschlager, Gunn & Hain. pp. 131-138.

Coppock, Rob. 1981. "Managing Technological Hazards" in: O'Riordan, T. & Turner, R. Kerry. Progress in Resource Management and Environmental Planning, Volume 3 . Chichester, England: John Wiley & Sons. pp. 203-217.

Cox, Robert W. & Jacobson, Harold K. 1973. The Anatomy of Influence. Decision Making in International Organizations . New Haven, New Jersey: Yale University Press.

Crandall, Robert W. & Lave, Lester B. 1979. The Scientific Basis of Health and Safety Regulations. Papers from a Conference . Washington, D.C.: Brookings Institute.

Damary, Roy & Hurst, Graham A.S. 1982. "A Study of Recall Practices Among Manufacturers of Consumer Products" The Geneva Papers on Risk and Insurance Vol.7, No.22 (January) pp. 27-66.

DeFinetti, B. 1937. "La prevision: ses lois logiques, ses sources subjectives" Annales de L'Institute Henri Poincare Vol. 7, pp. 1-68.

Dierkes, Meinolf. 1980. "Assessing Technological Risks and Benefits" in: Dierkes, Meinolf; Edwards, Sam & Coppock, Rob. (Eds.) Technological Risk: Its Perception and Handling in the European Community . Cambridge, Massachusetts: Oelgeschlager, Gunn & Hain. pp. 21-30.

Douglas, Mary & Wildavsky, Aaron. 1982. Risk and Culture; An Essay on the Selection of Technical and Environmental Dangers . Berkeley, California: University of California Press.

Drucker, Peter F. 1980. Managing in Turbulent Times . London: Heinemann.

Edwards, W.; Lindman, J. & Savage, L.J. 1968. "Baysian

Statistical Inference for Psychological Research" Psychological Review Vol. 70, pp. 193-242.

Eisenhardt, Giesela & Krebsbach, Camilla. 1978. Sterotype Wahrnehmung: Ihr Einfluss auf Interaktion und Kommunikation im Bereich politischer Plannung von grosstechnologische Einrichtungen . Frankfurt: Battelle-Institut (Bericht 400/3).

Emery, F.E. & Trist, E.L. 1965. "The Causal Texture of Organizational Environments" Human Relations Vol. 18,. pp. 21-32.

Encyclopaedia Britannica. 1981. "Safety Engineering" Encyclopaedia Britannica Vol.17. pp. 137-145.

Fischhoff, Baruch; Hohenemser, Christoph; Kasperson, Roger E. & Kates, Robert W. 1978. "Handling Hazards" Environment Vol.20, No.7 (September) pp. 16-20, 32-37.

Fischhoff, Baruch; Slovic, Paul & Lichtenstein, Sarah. 1978. "Fault Trees: Sensitivity of Estimated Failure Probabilities to Problem Representation" Journal of Experimental Psychology: Human Perception and Performance No. 4, pp. 342-355.

Fishbein, Martin & Ajzen, Icek. 1975. Belief, Attitude, Intention and Behavior: An Introduction to Theory and Research . Reaading, Massachusetts: Addison-Wessley.

Fishburn, P. 1970. Utility Theory for Decision Making . New York: Wiley.

Friedman, M. & Savage, L.J. 1948. "The Utility Analysis of Choices Involving Risks" Journal of Political Economy Vol. 56. pp. 279-304.

Grad, Frank P.; Rosenthal, Albert J.; Rockett, Laurie R.; Fay, James A.; Heywood, John; Dain, John F.; Ingram, Gregory & Tietenberg, Thomas. 1975. The Automobile, and the Regulation of Its Impact on the Environment . Norman, Oklahoma: University of Oklahoma Press.

Green, Kenneth & Hudspith, Robert. 1979. "Trade Unions and Technical Expertise--The Control of Asbestos Dust in British Workplaces" in: Skoie, Hans (Ed.) Scientific Expertise and the Public . Oslo: Institute for Studies in Research and Higher Education. pp. 153-170.

Green, M.R. & Serbein, O.N. 1978. Risk Management; Text and Cases . New York: Reston Publishing Co.

Hadden, Susan G. 1982. "Scientific Data and Public Policy" Science Vol. 216 (30 April) pp. 508-509.

Hammer, W. 1972. Handbook of System and Product Safety . Englewood Cliffs, New Jersey: Prentice-Hall.

Hanley, John W. 1981. "Monsanto's 'Early Warning' System" Harvard Business Review Vol. , No. (November/December) pp. 107-116.

Hardin, Garrett. 1968. "The Tragedy of the Commons" Science Vol. 162 (13 December) pp. 1244-1248.

Hays, S.P. 1959. Conservation and the Gospel of Efficiency . Cambridge, Massachusetts: Harvard University Press.

Head, George. 1982. "Continuing Evolution of the Risk Management Function and Education in the United States" The Geneva Papers on Risk and Insurance Vol.7, No.23 (April) pp. 180-186.

Hirschleifer, J. & Riley, John G. 1979. "The Analytics of Uncertainty and Information--An Expository Survey" Journal of Economic Literature Vol XVII (December) pp. 1375-1421.

Hofmann, Juergen. 1981. Erweiterte Nutzen-Kosten-Analyse . Goettingen, Germany: Vandenhoeck & Ruprecht.

Hofstadter, Douglas R. 1982. "About Two Kind of Inquiry" Scientific American Vol. 246, No. 2 (February) pp. 14-19.

Inglehart, Ronald. 1971. "The Silent Revolution in Europe: International Change in Post-Industrial Societies" American Political Science Review Vol. 65. pp. 991-1017.

Inglehart, Ronald. 1977. The Silent Revolution: Changing Values and Political Styles in Western Publics . Princeton, New Jersey: Princeton University Press.

Inglehart, Ronald. 1981. "Post-Materialism in an Environment of Insecurity" American Political Science Review Vol. 75, No. 4 pp. 1-22.

Irwin, George. 1978. Modern Cost-Benefit Methods. An Introduction to Financial, Economic and Social Appraisal of Development Projects. London: MacMillan.

Jungermann, Helmut; von Winterfeldt, Detlof & Coppock, Rob. 1982. Analysis, Evaluation and Acceptability of Hazardous Technologies and Their Risks. A Workshop Report. Berlin: International Institute for Environment and Society, DP-82-2.

Kahneman, Daniel & Tversky, Amos. 1973. "On the Psychology of Prediction" Psychological Review Vol. 80, No. 4 (July) pp. 237-251.

Kahneman, Daniel & Tversky, Amos. 1979. "Prospect Theory: An Analysis of Decisions under Risk" Econometrica Vol. 47, No. 2 (March) pp. 263-291.

Kates, Robert W. 1978. Risk Assessment of Environmental Hazard . SCOPE Report No. 8. Chichester, England: John Wiley & Sons.

Knoedgen, Gabriele. 1981. The STEAG Coal-Fired Power Plant at Voerde or Changing German Clean Air Act Policy. Berlin: International Institute for Environment and Society (IIUG/DP 81-11).

Kueppers, G. & Nowtney, H. (Eds.) 1979. The Impact of the Nuclear Controversy on Decision Making Structures . Bielefeld, West Germany: University of Bielefeld, Forschungsschwerpunkt Wissenschaftsforschung.

Leibenstein, Harvey. 1976. Beyond Economic Man, A New Foundation for Microeconomics . Cambridge, Massachusetts: Harvard University Press.

Lewis, Harold W. 1980. "The Safety of Fission Reactors" Scientific American Vol. 242, No. 3 (March) pp. 33-45.

Lindbloom, Charles E. 1977. Politics and Markets; The World's Political-Economic Systems . New York: Basic Books.

Lowrance, William W. 1976. Of Acceptable Risk. Science and the Determination of Safety . Los Altos, California: William Kaufmann, Inc.

Luce, R.Duncan & Raiffa, Howard. 1957. Games and Decisions: Introduction and Critical Survey . New York: Wiley.

Luft, Harold S. 1976. "Benefit-Cost Analysis and Public Policy Implementation: From Normative to Positive Analysis" Public Policy Vol. 24, No. 1, pp. 437-462.

Manville Corporation. 1982. Despite Strong Business, Litigation Forces Manville to File for Reorganization. International Hearld Tribune . Monday, 6 September, p. 15. (Advertisement)

Markowitz, Harry M. 1959. Portfolio Selection: Efficient Diversification of Investments . Cowles Foundation for Research in Economics, Yale University, Monograph 16. New York: Wiley.

Maslow, Abraham H. 1954. Motivation and Personality . New York: Harper.

Mathoeffer, Hans. (Ed.) 1977. Verbraucherforschung . Frankfurt: Umschau Verlag.

Mayer-Tasch, Peter Cornelius. 1976. Die Burgeriniativenbewegung. Der aktive Buerger als rechts- und politikwissenschaftsliches Problem. Reinbek bei Hamburg: Rowohlt Taschenbuch Verlag.

McConnel, G. 1965. "The Conservation Movement: Past and Present" in: Burton, I. & Kates, R.W. (Eds.) Resource Management and Conservation . Chicago: University of Chicago Press.

McCormick, Norman J. 1981. Reliability and Risk Analysis;

Methods and Nuclear Power Applications. New York: Academic Press.

Mc'Gonigle, R.Michael & Zacher, Mark W. 1979. Tankers at Sea. Berkeley, California: University of California Press.

Mishan, E.J. 1976. Cost-Benefit Analysis. New York: Praeger.

National Academy of Engineering. 1972. Perspectives on Benefit-Risk Decision Making. Washington, D.C.: The Academy.

Nelkin, Dorothy. 1968. "Threats and Promisses: Negotiating the Control of Research" Daedalus Vol. 107, No. 2 (Spring) pp. 557-583.

Nelkin, Dorothy. 1979. "Science, Technology and Political Conflict: Analyzing the Issues" in: Nelkin, Dorothy (Ed.) Controversy: Politics of Technical Decisions. New York: Sage Publications. pp. 9-22.

Nelkin, Dorothy & Pollak, Michael. 1980. "Consensus and Conflict Resolution" in: Dierkes, Meinolf; Edwards, Sam & Coppock, Rob. (Eds.) Technological Risk: Its Perception and Handling in the European Community. Cambridge, Massachusetts: Oelgeschlager, Gunn & Hain. pp. 65-75.

Neumann, J.von & Morgenstern, O. 1947. Theory of Games and Economic Behavior. Princeton, New Jersey: Princeton University Press.

Noll, Roger G. 1981. "Energy Data and Political Polarization" Science Vol. 215 (27 November) p. 1019.

Nowothy, Helga. 1979a. Kernenergie: Gefahr oder Notwendigkeit. Frankfurt: Surkamp.

Nowotny, Helga. 1979b. "Experts in a Participatory Experiment: The Austrian Debate on Nuclear Energy" in: Skoie, Hans (Ed.) Scientific Expertise and the Public. Oslo: Institute for Studies in Research and Higher Education. pp. 153-170.

Nuclear Regulatory Commission. 1975. Reactor Safety Study: An Assessment of Accident Risks in U.S. Commercial Nuclear Power Plants. Washington, D.C.: National Technical Information Service.

O.E.C.D. 1975. Methodological Guidelines for Social Assessment of Technology. Paris: Organization for Economic Cooperation and Development.

O.E.C.D. 1982. Procedings of the Workshop on the Control of Exising Chemicals Under the Patronage of the Organization for Economic Cooperation and Development. Berlin: Umweltbundesamt.

Olson, Douglas B. & Simkiss, John A. Jr. 1982. "An Overview of Risk Management" The Geneva Papers on Risk and Insurance Vol.7, No.23 (April) pp. 114-128.

Olson, Mancur. 1971. The Logic of Collective Action . Cambridge, Massachusetts: Harvard University Press.

O'Riordan, Timothy. 1979. "Environmental Impact Analysis and Risk Assessment in a Management Perspective" in: Goodman, G.T. & Rowe, W.D. (Eds.) Energy Risk Management . London: Academic Press. pp. 21-36.

O'Riordan, Timothy. 1981. Environmentalism . London: Pion, Ltd.

Otway, Harry J. 1980. "The Perception of Technological Risks: A Psychological Perspective" in: Dierkes, Meinolf; Edwards, Sam & Coppock, Rob. (Eds.) Technological Risk: Its Perception and Handling in the European Community . Cambridge, Massachusetts: Oelgeschlager, Gunn & Hain. pp. 35-44.

Otway, Harry & Fishbein, Martin. 1977. Public Attitudes and Decision Making. Laxenburg, Austria: International Institute for Applied Systems Analysis (IIASA-RM-77-54).

Otway, Harry & von Winterfeldt, Detlof. 1982. "Beyond Acceptable Risk: On the Social Acceptability of Technologies" Policy Sciences Vol. 14, No. 3, pp. 247-256.

Page, Talbot. 1978. "A Generic View of Toxic Chemicals and Similar Risks" Ecology Law Quarterly Vol. 7, No. 2. pp. 207-244.

Pappi, F.U. & Laumann, E.O. 1974. "Gesellschaftliche Wertorientierungen und politisches Verhalten" Zeitschrift fuer Soziologie Vol. 3, No. 2 (April) pp. 157-188.

Pearce, David W. 1980. "The Preconditions for Achieving Consensus in the Context of Technological Risk" in: Dierkes, Meinolf; Edwards, Sam & Coppock, Rob. (Eds.) Technological Risk: Its Perception and Handling in the European Community . Cambridge, Massachusetts: Oelgeschlager, Gunn & Hain. pp.57-63.

Pfeffer, Jeffrey & Salancik, Gerald R. 1978. The External Control of Organizations--A Resource Dependence Perspective. New York: Harper & Row.

Porter, Alan L.; Rossini, Frederick A.; Carpenter, Stanley R; Roper, A.T.; Larsen, Ronal & Tiller, Jeffery S. 1980. A Guidebook for Technology Assessment and Impact Analysis . New York: North Holland.

Posner, Richard A. 1977. Economic Analysis of Law Second Edition. Boston: Little, Brown & Company.

Raiffa, Howard. 1968. Decision Analysis: Introductory

Lectures on Choices under Uncertainty . Reading, Masachusetts: Addison-Wesley.

Ravetz, Jerome R. 1980. "Public Perceptions of Acceptable Risk as Evidence of Their Cognitive, Technical and Social Structure" in: Dierkes, Meinolf; Edwards, Sam & Coppock, Rob. Technological Risk: Its Perception and Handling in the European Community . Cambridge, Massachusetts: Oelgeschlager, Gunn & Hain. pp. 35-44.

Renn, O. 1981a. Wahrnehmung und Akzeptanz technischer Risiken, Band 1 . Juelich: Kernforschungsanlage Juelich (Juel-Spez-97 Bd I).

Renn, O. 1981b. Wahrnehmung und Akzeptanz technischer Risiken, Band 2 . Juelich: Kernforschungsanlage Juelich (Juel-Spez-97 Bd II).

Rock, Reinhard; Biervert, Bernd & Fischer-Winkelmann, Wolf, F. 1980. "A Critique of Some Fundamental Theoretical and Practical Tenets of Present Consumer Policy" Zeitschrift fuer Verbraucherpolitik Vol.4, No.2, pp. 93-101.

Ronge, Volker & Schmieg, Guenther. 1972. Restriktionen politischer Planung . Frankfurt: Athenaeum Verlag.

Rowe, William D. 1977. An Anatomy of Risk . New York: John Wiley & Sons.

Ruggie, John G. 1975. "International Responses to Technology: Concepts and Trends" International Organization Vol. 29, No. 3. pp. 557-583.

Sandler, Todd & Tschirhart, John T. 1980. "The Economic Theory of Clubs: An Evaluative Surfey" Journal of Economic Literature Vol. 18, No. 4 (December) pp. 1481-1521.

Sassone, Peter G. & Schaffer, Wiliam A. 1978. Cost-Benefit Analysis--A Handbook . New York: Academic Press.

Savage, L.J. 1954. The Foundations of Statistics . New York: Wiley.

Schnee, Jerome E. 1979. "Regulation and Innovation: U.S. Pharmaceutical Industry" California Management Review Vol. XXII, No. 1 (Fall) pp. 23-32.

Selznick, P. 1957. Leadership in Administration . Evanston, Illinois: Row Peterson.

Shooman, M.L. 1968. Probabilistic Realiability: An Engineering Approach . New York: McGraw-Hill.

Sibthorp, M.M. 1975. The North Sea; Challange and Opportunity . London: Europe Publications.

Simon, Herbert A. 1957. Models of Man . New York: Wiley.

Skolnikoff, Eugene G. 1972. The International Imperatives of Technology. Research Series No. 16. Berkley, California: Institute of International Studies, University of California.

Slovic, Paul; Fischhoff, Baruch & Lichtenstein, Sarah. 1980. "Facts and Fears: Understanding Perceived Risk" in: Schwing, Richard C. & Albers, Walter A. Jr. (Eds.) Societal Risk Assessment. How Safe is Safe Enough? New York: Plenum Press. pp. 181-214.

Smith, C.O. 1976. Introduction to Reliability in Design. New York: McGraw-Hill.

Starr, Chauncey. 1969. "Social Benefit versus Technoogical Risk" Science Vol.165 (19 September) pp. 1232-1238.

Stokey, E. & Steckhauser, R. 1978. A Primer for Policy Analysis. New York: Norton.

Thomas, K.; Maurer, D.; Fishbein, M. & Otway, H.J. A Comparative Study of Public Beliefs About Five Energy Systems. Laxenburg, Austria: International Institute for Applied Systems Analysis (IIASA-RR-80-15).

Thomas, K.; Swaton, E.; Fishbein, M. & Otway, H.J. Nuclear Energy: The Accuracy of Policy Makers. Laxenburg, Austria: International Institute for Applied Systems Analysis (IIASA-RR-80-18).

Thompson, Michael. 1980a. Political Culture: An Introduction. Laxenburg, Austria: International Institute for Applied Systems Analysis. (WP-80-175)

Thompson, Michael. 1980b. The Social Landscape of Poverty. Laxenburg, Austria: International Institute for Applied Systems Analysis. (WP-80-174)

Thompson, Michael. 1980c. An Outline of the Cultural Theory of Risk. Laxenburg, Austria: Internationaal Institute for Applied Systems Analysis. (WP-80-1770

Touraine, Alaine. 1971. The Post Industrial Society; Tomorrow's Social History: Classes, Conflicts and Culture in the Programmed Society. Translated by Leonard F.X. Mayhew. London: Wildwood House Ltd.

Tucker, William. 1977. "Environmentalism and the Liesure Class" Harper's Vol. 255 (December) pp. 49-56, 73-80.

Tversky, Amos & Kahneman, Daniel. 1974. "Judgment under Uncertainty: Heuristics and Biases" Science Vol. 185 (27 September) pp. 1124-1131.

Tversky, Amos & Kahneman, Daniel. 1981. "The Framing of Decisions and the Psychology of Choice" Science Vol. 211 (30 January) pp. 453-458.

Vleck, Charles & Stallen, Peter-Jan. 1980. "Rational and Personal Aspects of Risk" Acta Psychologica Vol. 45, pp. 273-300.

Vleck, Charles & Stallen, Peter-Jan. 1981. "Judging Risks and Benefits in the Small and in the Large" Organizational Behavior and Human Performance Vol. 28, pp. 235-271.

Waddington, J.I. 1982. "Report on the programme on European co-operation on environmetal health aspects on the control of chemicals" in: Umweltbundesamt. Proceedings of the Workshop on the Control of Existing Chemicals Under the Patronage of the Organization for Economic Co-operation and Development . Berlin: Umweltbundesamt. pp.45-82.

Wildavsky, Aaron. 1979. The Politics of the Budgetary Process . Third Edition. Boston: Little, Brown & Company.

Wildavsky, Aaron & Tenenbaum, Ellen. 1981. The Politics of Mis-trust. Estimating American Oil and Gas Resources . Beverly Hills, California: Sage.

Williams, Robert C. 1980. "Three Mile Island as History" Washington University Magazine (Fall) pp. 56-63.

Wynn, Brian. 1980. "Technology, Risk and Participation: On the Social Treatment of Uncertainty" in: Conrad, J. (Ed.) Society, Technology and Risk Assessment . London: Academic Press. pp. 174-208.

THE AUTHOR

After receiving degrees in physical science (BS) and public administration (MPA), Mr. Coppock joined the staff of Battelle Seattle Research Center in 1969, where he participated in research on cultural regions of the United States, corporate social accounting, and the scientific analysis of problems on the worldwide level. In 1973, Mr. Coppock became a member of the staff of Battelle-Institut e.V. in Frankfurt, West Germany and contributed to research on the relation between society and business, corporate social reporting, and social forecasting. Since 1976, he has been a Research Fellow at the International Institute for Environment and Society of the Science Center Berlin. There, Mr. Coppock has been involved in studies on risk assessment and management, social science research policy, and the social and environmental consequences of of tourism in the mountain regions of Nepal. Mr. Coppock is currently coordinator for research in West Germany contributing to the multi-year, international program on "The Future of the Autombile" organized by Massachusetts Institute of Technology. In addition, he coordinated the activities of this program which focus on the level of the European Economic Community. In 1983, Mr. Coppock completed the requirements for a Doctorate in Economics at the University of Wuppertal, in West Germany. He is author of numerous articles and research reports, and co-editor of Technological Risk: Its Perception and Handling in the European Community. Mr. Coppock is a member of the Editorial Board of the journal Risk Analysis. He has experience as a semi-professional cinematographer, photographer, and writer, and served as equipment manager and logistics officer for a mountaineering expidition in South America.